T0146286

Personal Identity and Fractured Selves

Personal Identity and Fractured Selves

Perspectives from Philosophy, Ethics, and Neuroscience

Edited by

Debra J. H. Mathews, Ph.D., M.A.

Hilary Bok, Ph.D.

Peter V. Rabins, M.D., M.P.H.

Johns Hopkins Berman Institute of Bioethics
Baltimore, Maryland

The Johns Hopkins University Press
Baltimore

Printed in the United States of America on acid-free paper
9 8 7 6 5 4 3 2 1

The Johns Hopkins University Press
2715 North Charles Street
Baltimore, Maryland 21218-4363
www.press.jhu.edu

Library of Congress Cataloging-in-Publication Data
Personal identity and fractured selves : perspectives from philosophy, ethics, and neuroscience / edited by Debra J. H. Mathews, Hilary Bok, and Peter V. Rabins.
p. cm.
Includes bibliographical references and index.
ISBN-13: 978-0-8018-9338-4 (hardcover : alk. paper)
ISBN-10: 0-8018-9338-0 (hardcover : alk. paper)
1. Identity (Psychology). 2. Personality disorders. 3. Neuropsychology. I. Mathews, Debra J. H. II. Bok, Hilary, 1959– III. Rabins, Peter V.
[DNLM: 1. Self Concept. 2. Brain—physiopathology. 3. Neurosciences. 4. Personal Autonomy. 5. Personality. 6. Philosophy. BF 697 P467 2009]
BF697.P467 2009
155.2—dc22 2008050543

A catalog record for this book is available from the British Library.

The Johns Hopkins University Press uses environmentally friendly book materials, including recycled text paper that is composed of at least 30 percent post-consumer waste, whenever possible. All of our book papers are acid-free, and our jackets and covers are printed on paper with recycled content.

Contents

Contributors

Samuel Barondes, M.D., Professor, Department of Psychiatry, Jeanne and Sanford Robertson Endowed Chair in Neurobiology and Psychiatry, University of California, San Francisco, San Francisco, California

David M. Blass, M.D., Assistant Professor, Department of Psychiatry and Behavioral Sciences, Johns Hopkins University School of Medicine, Baltimore, Maryland

Hilary Bok, Ph.D., Henry R. Luce Professor of Bioethics and Moral and Political Theory, Department of Philosophy, Johns Hopkins School of Arts and Sciences, Baltimore, Maryland

Patrick Duggan, A.B., Senior Research Program Coordinator, Johns Hopkins Berman Institute of Bioethics, Baltimore, Maryland

Michael S. Gazzaniga, Ph.D., Director, SAGE Center for the Study of Mind, University of California, Santa Barbara; Member, President's Council on Bioethics

Debra J. H. Mathews, Ph.D., M.A., Assistant Director for Science Programs, Johns Hopkins Berman Institute of Bioethics, Baltimore, Maryland

John Perry, Ph.D., Henry Waldgrave Stuart Professor of Philosophy, Stanford University, Palo Alto, California

Peter V. Rabins, M.D., M.P.H., Professor of Psychiatry and Behavioral Sciences, Co-Director, Division of Geriatric Psychiatry and Neuropsychiatry, Johns Hopkins Hospital, Baltimore, Maryland

Alan Regenberg, M.Be., Senior Research Program Coordinator, Johns Hopkins Berman Institute of Bioethics, Baltimore, Maryland

Carol Rovane, Ph.D., Professor of Philosophy, Chair, Department of Philosophy, Columbia University, New York, New York

Marya Schechtman, Ph.D., Professor of Philosophy, University of Illinois at Chicago, Chicago, Illinois

Maura Tumulty, Ph.D., Assistant Professor of Philosophy, Colgate University, Hamilton, New York

Preface

What makes me *me*, and not you? Am I still *me*, or the same *me*, no matter how many changes occur in my internal and external life? These are questions that philosophers consider and have constructed theories of personal identity to resolve. These are also questions that clinical specialists in neurology, psychiatry, psychology, and neurosurgery regularly confront in the care of patients with a variety of brain disorders and injuries and that neuroscientists are increasingly poised to explore. Yet philosophers, medical specialists, and neuroscientists rarely engage together around the difficult conceptual and ethical challenges of understanding and giving moral meaning to personal identity. This volume represents an initial attempt to connect the different standpoints and ways of knowing that these different disciplines represent. Our working assumptions are that philosophical theories of personal identity will be enriched and advanced by attempts to grapple with the specific realities of particular clinical cases and that clinicians and neuroscientists will benefit from a richer understanding of how to view questions of personal identity in philosophical terms.

With support from the Greenwall and Dana Foundations, we asked three prominent philosophers with competing accounts of personal identity to respond to four clinical case studies that were designed to pose different challenges to intuitions about personal identity. We also asked two renowned neuroscientists to respond to the philosophers on the basis of their clinical and research experience. This book is the product of that endeavor.

There are reasons that this book, with its interdisciplinary focus, is rare in the literature on personal identity. Philosophers, neuroscientists, and clinicians not only think about personal identity in different ways but also speak different languages. The introduction to this volume includes a guide for translating between philosophy and neuroscience, as well as overviews of how both philosophers and neurobiologists think about personal identity. Part I addresses these language and conceptual barriers head on; it also includes the four cases that were presented to the philosophers: Alzheimer disease, frontotemporal dementia, deep brain stimulation, and steroid psychosis. The philosophers' analyses of the cases in the context of their theories of personal identity are presented in part II, and the neuroscientists' responses to the philosophers are presented in part III. The conclusion summarizes what is common to and what separates the differing disciplinary perspectives represented in this volume and also summarizes what this project achieved.

Guy M. McKhann, M.D.
Professor of Neurology and Neuroscience,
Department of Neurology,
Johns Hopkins University School of Medicine

Ruth R. Faden, Ph.D., M.P.H.
Philip Franklin Wagley Professor of Biomedical
Ethics and Executive Director, Johns Hopkins
Berman Institute of Bioethics; Professor,
Department of Health Policy and Management,
Johns Hopkins University
Bloomberg School of Public Health;
Professor, Department of Medicine,
Johns Hopkins University School of Medicine

Personal Identity and Fractured Selves

INTRODUCTION

A "Two Cultures" Phrasebook

Debra J. H. Mathews, Ph.D., M.A., Alan Regenberg, M.Be., and Patrick Duggan, A.B.

This volume is the product of a symposium convened by the Johns Hopkins Berman Institute of Bioethics Program in Ethics and Brain Sciences. The project brought together prominent philosophers and neuroscientists to address the following question: When an individual's personality changes radically, as a consequence of either disease or intervention, should this changed individual still be treated as the same person? Over the course of the symposium, it became clear that different understandings of terms such as *person, personhood,* and *self* had a significant influence on the discussion. The varying usage across disciplines caused initial misunderstandings among symposium speakers and participants, frustrating productive conversation. For example, in philosophy, many of these terms are technical, though they are common colloquially. To move beyond these lexical differences and thus allow us to address the conceptually driven similarities and differences between the neuroscientists' and philosophers' responses to the issues of personal identity raised by the four case studies, we first spend some time discussing these terms. We describe how the various speakers used the words that proved to be most open to interpretation or differing technical and colloquial uses.

Person

The concepts of *personal identity* and *changes in personal identity* are premised on certain notions of what it is to be a *person*. Without attending to the different ways in which people understand this word, it is easy to become mired in disagreements about who or what "counts as" a person, which, interesting though they are, preclude meaningful discussion about *personal identity*, the main topic of the symposium. This question of what it means to be a person is familiar to most philosophers and ethicists from debates about moral status and personhood. Scientists and clinicians, however, grapple with this (largely theoretical) question less often, and when they do, they tend to answer it in ways quite different from those of philosophers.

Each philosopher's account of personal identity rests on a nuanced conception of what it means to be a person. Marya Schechtman consistently uses the term *forensic person* rather than simply *person*. In so doing, she makes it clear that she is not trying to overturn the typical notion of a person in everyday usage. Rather, she focuses on the properties that make individuals morally responsible agents who stand in relation to other morally responsible agents. Hence, her account characterizes a person as someone with specific capacities (e.g., the ability to enter into moral and contractual commitments and to form a narrative conception of himself or herself). An important implication of this is that forensic personhood admits of degrees. Individuals with advanced dementia, who have limited narrative conceptions of themselves, possess forensic personhood to a lesser degree than a healthy individual. Those who are entirely incapacitated are not forensic persons.

John Perry characterizes persons as "systems that pick up information from experience, develop and sustain goals, and apply the information to achieve their goals." To a first approximation, human beings with human bodies count as persons, but our experiences with special cases may call this approximation into question. We may encounter human bodies that do not perceive or respond to the world around them or that seem incapable of forming goals and acting on them. We cannot expect that such individuals will be able to state their preferences, to make and pursue goals, or to do any number of things that we expect persons to be able to do. As such, we cannot reasonably hold them responsible for what they do or fail to do. Perry asserts that, on thorough investigation, we may be forced to conclude that such individuals are not persons, although we may still regard them as diminished and fractured versions of the persons they once were.

In these two accounts, we can see a movement away from what most people take as a general rule—namely, that there is normally a direct correspondence between persons and living human beings. While Schechtman and Perry begin from this assumption and show how it cannot escape careful scrutiny, Carol Rovane rejects it from the outset. In her account, there is no assumed connection between human bodies and persons. Persons are rational agents committed to the goal of achieving overall rational unity. It *could* be the case that many, even most, human beings are persons, but it is decidedly not Rovane's aim to connect the idea of being human with being a person. Things that are larger than human beings (teams, university departments, corporations) or smaller than human beings (facets of individuals committed to achieving overall rational unity within themselves) may count as persons more clearly than do individual human beings.

Neuroscientist Michael Gazzaniga seeks some common ground with the philosophers in his contention that "what makes us persons, rather than merely creatures, is our ability to create a story about ourselves." Something unique emerges out of the human body and human biology—something much like Schechtman's "narrative self-constitution"—that transforms a biological creature into a "person." Interestingly, Gazzaniga goes another step and locates this ability to create a story about oneself in a particular region of the brain, in what cognitive neuroscientists dub the "interpreter" system. This step is significant. While his account seems consonant with those of the philosophers, it still resists the idea that a person could exist in the absence of a human body. He agrees that persons are more than mere human bodies, but seems to argue that having a human body (or at least a human brain) is a necessary part of being a person.

Samuel Barondes steers clear of this question about who or what counts as a person, presumably subscribing to what we might call a commonsense view (with *commonsense* here referring to how the word *person* is employed in everyday language, not implying that this view is better or worse than some other view). He tends to use the term *personality* (see below) more often than *person*. Implicit in the commonsense view is the idea that stipulating criteria for personhood either does not help our discussion (because we still have to address the question of how to regard individuals who are not persons) or, worse, hinders our discussion (because it hearkens back to situations in which entire groups of people were mistreated on the grounds that they did not count as persons). For example, even if it were possible to conclude that an individual in a vegetative state or with advanced dementia is not a person (or is less a person than we are), doctors would still have to know how to care for the individual properly, and this is the question that matters for clinicians.

These concerns are addressed by the philosophers (Schechtman and Rovane), who assert that how we treat individuals need not depend on whether they are persons or not. For the philosophers, the question of personhood is more theoretical than practical. We need a conception of *person* if we hope to talk coherently about personal identity. How we treat individuals is a separate (albeit related) question that need not necessarily (though it may) rest on the question of personhood. Of course, this line of argument shows that we do not *need* to employ the concepts of person and personal identity in neuroscience and medicine to do the work of neuroscience and medicine, but elucidating these concepts may shed light on certain questions that we encounter in those fields.

Same Person / Different Person Criteria for Personal Identity

Intimately linked to the concept of person and personhood is the idea that a person may change over time into a "different person." While this is something that we often say colloquially (e.g., she's not acting like herself; he's a different person when he's around her; he's not the same person he used to be), in academic philosophy, the idea of different/changing persons can be a highly specified technical concept. As with diverse understandings of the word *person*, the different ways in which scholars at the symposium understood and used the concept of *same* and *different* persons caused some debate. This concept—that human bodies and human persons can diverge—is at the heart of the various contributors' interpretations of the clinical cases.

Schechtman states that "the limits of a forensic person are set by the limits of a narrative. As long as a single narrative continues there is a single forensic person." This *narrative self-constitution view* operates under at least two constraints: that the human being is able to articulate relevant aspects of her self-narrative (e.g., that she has children and works in a university; why she enjoys rock climbing but not running; what she plans to do after retirement) and that her narrative "fundamentally cohere[s] with reality" (i.e., the self-narrative she articulates is factually accurate or, at least, is open to revision when faced with evidence of error). If either of these constraints is violated, the individual is not fully a forensic person. If an individual cannot construct a single, coherent narrative, with a past, present, and future that are continuous and related, less *or more* than a single forensic person may inhabit the human body; that is, a different person may come to exist within a given human body.

In contrast, Perry rejects the language of "different persons" for the four case studies. He writes: "We are dealing with persons . . . who, as the disease progresses, have diminished selves . . . But we are not dealing with nonpersons or different persons, in the strong sense. Their identities change, in the psychological sense, but they are the same person, diminished and changed." However, he seems to leave open the possibility that there are cases in which, if a person is not changed, she, at a minimum, ceases to be a person while still alive.

As described above, Rovane rejects the notion that persons and human bodies must exist in a one-to-one ratio. For her, "*the existence of a person . . . is always bound up with the exercise of effort and will*" (italics hers) and a person is a rational agent committed to achieving overall rational unity within/across her life. Continuing to be the *same* person, or becoming a *different* person, then, depends on the endurance of one's commitments over time. A discontinuity of *commitments,* such as marriage or ardent environmentalism, versus *dispositions,* such as a preference for chocolate over pistachio ice cream, is likely to result in the emergence of a different person. The loss of commitments results in the disappearance of a person.

Gazzaniga does not discuss the idea of same or different persons. However, he does grant that changes in the brain or changes in the environment may result in changes to one's narrative—how it is constructed and how it is understood.

Barondes's view is that humans/persons are complex creatures who "show different faces of ourselves in different contexts," and that when we behave differently (e.g., at work and at home), we are treated differently, accordingly. In his view, this does not imply that we are different persons at work and at home, but indicates that we are "the same person—replete with the inconsistencies and contingent behaviors and the unconscious motives and the self-deceptions that each of us has in considerable measure." Further, in the case of human beings with neurological conditions of various sorts, Barondes likewise dismisses the idea that a human being becomes a different person over time.

Personality

Another term central to the concept of personal identity that is frequently used but not often well defined is *personality*. *Dorland's Illustrated Medical Dictionary*, thirtieth edition, defines personality as "the characteristic way that a person thinks, feels, and behaves; the relatively stable and predictable part of a person's thought and behavior; it includes conscious attitudes, values, and styles as well as unconscious conflicts and defense mechanisms." All three philosophers and one of the neuroscientists make some use of the term *personality*. These usages seem, prima facie, to be fairly consistent, yet closer examination reveals some potentially important differences that highlight the unique ways in which each author approaches the problem of personal identity. Further, the context in which *personality* is used may provide clues to which of the similarities and differences in the accounts can be attributed more to lexical hurdles than to true conceptual barriers.

Schechtman first uses *personality* in relation to the rare condition known as dissociative identity (or "multiple personality") disorder (a disorder also used as an example by the other philosophers), as this condition provides a convenient challenge to concepts of personal identity. She writes: "we are asking, among other things, whether these personalities should be treated as separate persons for the practical purposes described." Thus we can see that Schechtman is considering *personality* to be distinct from yet related to *person*. Schechtman also uses *personality* in the context of her discussion of the case studies. She mentions how the cases describe personality changes and their implications for forensic personhood. In this way, she describes *personality* as a relatively stable trait that is strongly related yet distinct from personal identity. In Schechtman's analysis, forensic personhood can tolerate some amount of personality change.

Perry gets to *personality* by way of a notion he calls *identity in the psychological sense*. "When a psychologist or an ordinary individual (i.e., not a philosopher) talks about the identity of a person, he or she mainly has in mind not something that could be decided by fingerprints or a driver's license picture, but an enduring structure within the person, his or her own individual combination of beliefs, goals, habits, and traits of character and personality, the pattern that, as we might say, *makes* the person who he or she is." He notes that personality is a trait included among the qualities that encompass personal identity. Perry also discusses personality with respect to the movie *The Three Faces of Eve,* about which he writes: "Here we have one human, but three 'personalities.'"

Rovane uses *personality* in the context of the changes that she sees as important factors in analyzing the clinical cases. It is not the personality change *itself* that would, in her view, threaten personal identity but, rather, that personality changes may *lead persons* to "obliterate old commitments and usher in new commitments in their place and that this may mark the end of one person's life and the beginning of a new one—a nonbiological death and birth, as it were." So here, yet again, we see personality as one among the constellation of features that comprise a person.

Interestingly, *personality* does not appear in Gazzaniga's contribution, and his accounts of the cases seem to describe personality mechanistically. Describing the importance of the proper functioning of various subsystems within the brain and the implications for the human organism when these systems malfunction, Gazzaniga writes, "It is clear that the sense of self can be compromised by any number of diseases or circumstances. Suppose the 'self system' doesn't light up, or suppose it lights up but the patient shows no behavioral sense of self. Suppose the medical community then says: 'There's no sense of self; there's nothing there. We have a theory about this person in our minds, but the person doesn't have a theory about himself in his own mind.'" Curiously, within this set of systems, Gazzaniga's descriptions seem to personify portions of the brain (e.g., the interpreter). So, in his account, we seem to have at least two entities, the interpreter and the volitional self, engaged in some form of dialogue about the nature of reality. This is all described in a unique language that does not appear in any of the other accounts. Thus, the question remains, if Gazzaniga were asked to define *personality* within his account, would it yield an account that is similar to the others?

As a psychiatrist, it is not surprising that Barondes explicitly defines *personality* in terms that mirror the medical definition given above. He defines it as "the characteristic pattern of thoughts, feelings and behaviors of an individual." He goes on to distinguish between a trait such as personality and the more volatile sorts of states reflected in cognitive performance or global functioning, such as might be measured on the Global Assessment of Functioning

(GAF) scale in the *Diagnostic and Statistical Manual of Mental Disorders,* fourth edition. In Barondes's account, personality traits are described as relatively stable and enduring throughout adulthood, which is an understanding similar to Schechtman's and Perry's; however, these traits can be threatened and changed by disease. For Barondes, personality is a distinct notion, separate from personal identity, for the challenges presented by the medical case studies. He presents the central question of the symposium, with which we opened this introduction: "When an individual's personality changes radically, as a consequence of either disease or intervention, should this individual still be treated as the same person?" He rejects the notion that we should ever consider the patients in cases such as the four examples as anything other than the "same person" who was there originally: "patients have so much in common with who they were in their premorbid state that, in my view, there is no compelling reason to think of them as being anything other than some version of the same person." Instead, Barondes would prefer to consider these patients as diseased, disordered, and poorly functioning versions of the same person. In his view, then, pathological processes can mask or change an individual's personality, but these changes are best seen as deviations from the premorbid state.

Thus, while Barondes seems to identify *personality* as a term to describe the overall pattern of identity, Perry offers *personality* as being among the elements that constitute identity. While this seems to be a subtle difference, it is important. It could indicate that Barondes's objection to considering the patients in the case studies as anything but the same person is really an objection to the terminology being considered by the philosophers. Of note, the change in "forensic personhood" considered by Schechtman may not threaten Barondes's assertion that an individual with impaired cognitive functioning and personality changes is still the same person.

Self/Selves

Self, or *selves,* is yet another term used in reference to human beings (though again, not necessarily in a one-to-one ratio with human bodies). When we speak of "the self," we are often speaking not of the world's perception of a particular human being but of a specific human's sense of her own being. We use the term in compounds as *self-determination, self-esteem, self-respect,* and so forth, all to describe a human's relationship to her own being.

Self is indeed closely related to *person*, but it is distinct and still open to interpretation by philosophers and scientists.

While Schechtman does not define *self* explicitly (a trend among the contributors), she nonetheless uses the word in her discussion. In her chapter, *self* is used exclusively in compound words (herself, oneself, self-narrative, self-constitution, self-determination, etc.). She seems to operate under a definition of *self* that requires the being for whom the term is used to be a forensic person but that does not equate *self* with *forensic person.*

Perry seems to have a similar definition, but he is explicit: "a self is just a person, thought of under the relation of identity . . . The basic concept of self is not of a special kind of object, but of a special kind of concept that we each have of ourselves." And this *self-notion* that each person has is key to Perry's person theory. So, again, a self can exist only when a person exists. Further, each person is a multiply centered self; that is, each of us has a range of goals and beliefs that struggle for ascendancy at any given time.

Like Schechtman, Rovane mainly uses the term *self* in compounds and does not define it explicitly. Like the other philosophers, Rovane seems to view *self* as traveling with but not identical to *person.* For example, she writes: "we cannot think of the right to *self*-determination as extending to times after competence has been lost unless we think of *the life of the self* (i.e., the person) as extending to those times as well."

In contrast to the other contributors, Gazzaniga has both a biological and a philosophical view of *self.* He writes: "the self is a narrator—sometimes an unreliable narrator—that tries to piece together a coherent story, to find some rational unity, and to give itself a feeling of control. The self has a theory about itself, a *person theory* that applies to others as well as itself." Furthermore, as described above, he locates this narrator in the brain. For Gazzaniga, the relationship between the *self* and a *person* seems to be inverted from the understandings of the other authors. For him, the person emerges from the self, whereas for the others, a self cannot exist without a person, or it emerges simultaneously with the person.

Barondes seems to be more in line with the philosophers on *self.* While he does not define the term, he probably would agree with Perry's definition, and he uses *self* in compounds, mostly when talking about and agreeing with Perry's views on the cases.

Mind versus Brain

The philosophers and neuroscientists also seem to be at odds with respect to the relationship (if any) of the mind to the brain. Of course, no one plausibly denies that the brain is a biological organ with certain anatomical and physiological features—precisely because the brain is a physical entity than can be measured, imaged, medicated, and operated on, there is no meaningful dispute about *what* it is. Gazzaniga lays out an excellent working description of the brain: "The brain is there to make decisions. It makes decisions on a second-by-second, moment-by-moment, day-to-day, week-to-week basis, and it accumulates information from all kinds of sources, putting the data into some kind of decision mechanism or decision network."

However, there *is* variation in how the contributors deploy terms for nonphysical analogues of the brain, most notably, *mind*. Among the philosophers, only Perry uses the term in anything other than a colloquial manner (such as "to have something in mind"). In his view, the brain and the mind either are identical (the brain *is* the mind) or are causally related (states of the brain can be said to *determine* mental states). The *mind*, and more ultimately the brain, is where our thoughts and ideas about the world reside. Coupled with the "decision mechanisms" in the brain that Gazzaniga speaks of, the brain connects our ideas about the world with other internal states (desires, goals, emotions, etc.) to coordinate our actions in some rational way. That is, we can connect apparent facts about the world with our own goals and then act in a way that might serve to advance those goals. Hence Perry steers clear of the Cartesian dualism in which the *mind* occupies a privileged position separate from the body.

But limiting ourselves to the terms *mind* and *brain* is perhaps too restrictive. The more substantive questions here are whether, how, and to what extent physical characteristics are bound up with the supposedly nonphysical (or psychological) aspects of who we are. Most contemporary accounts of personal identity (and all of the accounts given in this volume) are associated with the psychological characteristics of persons—memories, beliefs, wills, commitments, self-narratives, and the like. In each account, some constellation of these features and/or other related concepts is regarded as necessary and sufficient for establishing whether a person remains the same person over time. The philosophers remain skeptical that having recourse to physical criteria will tell us anything helpful that we could not otherwise establish using psychological criteria.

Schechtman's account leans heavily on the psychological capacity to develop a narrative self-conception, but she is silent about whether this capacity has a defined biological basis. Gazzaniga, as noted above, explicitly ties this capacity to the left hemisphere's interpreter system. This system is what enables an individual to make sense of the world and his or her place in it, to generate (post hoc) explanations for why events occurred the way they occurred. It is not clear whether Schechtman would endorse this view as a complete account of narrative self-constitution. More likely, she might grant that it contributes to narrative self-constitution but that other psychological capacities, such as memory, emotion, and the ability to deliberate about the future, also play a necessary role.

Rovane's skepticism most likely stems from a concern that focusing on physical criteria will lead us to believe (mistakenly, in her view) that persons are bound to individual human bodies, when, in fact, *commitment* and *will* can and do constitute persons of all different sizes. Whether or not they have any physical or biological basis is irrelevant, because these capacities are still capable of constituting persons that reside outside the bounds of an individual human body.

However, the skepticism of the philosophers regarding the importance of physical criteria does not eliminate the question of whether the psychological properties constitutive of personal identity are simply physical properties in disguise (see philosopher Maura Tumulty's chapter for a fuller discussion and one possible answer to this question). Stated another way, the more we understand about the structure and function of the brain, the more we realize that the mysterious and ephemeral psychological properties that we cherish may simply be the products of brain activity. If neuroscientists could locate *all* of the necessary aspects of personal identity in the nervous system, this would seem to suggest a more materialist, biological account of personal identity. Of course, the day when this claim can be either proven or refuted is a long way off, and so the philosophers' accounts are relatively safe and will be for some time.

Still, the question is worth asking for another reason. If we plan to apply questions about personal identity to clinical cases (as we set out to do), we need a way to connect what we know about the brain (e.g., the pathological processes involved in frontotemporal dementia or the mechanisms by which deep brain stimulation works in treating brain disorders) with the other changes (e.g., in memory, cognition, and movement) we can observe in persons. Though we may still wish to speak of psychological properties that cannot be fully explained in biological terms, we need to develop an account of how advances in neuroscience either do or do not affect our theories about personal identity. Presumably, it matters that we can attribute changes in a person's behavior to something happening in that person's brain, rather than attributing them to another, less concrete cause.

This is where philosophy and neuroscience can inform each other. If we know (which we do) that Alzheimer disease is irreversible and that it occurs irrespective of personal choice, this will inform our theories about personal identity for such patients. Conversely, if we know that the psychosis resulting from a choice to take steroids is different in many respects from other psychotic disorders that are not the result of a personal choice and are not reversible, this will have implications for how we think about personal identity in such cases. All of the philosophers' contributions in this volume bear these neuroscientific facts in mind when deliberating about the case studies. Hence, philosophical theories of personal identity have much to draw from neuroscience, without thereby having to submit to the notion that personal identity is reducible to biological or bodily identity.

PART I / Foundations

CHAPTER ONE

How Philosophers Think about Persons, Personal Identity, and the Self

Maura Tumulty, Ph.D.

Most people develop a sense of self early in their lives and think of them selves as persistent. Whatever dramatic personal changes they may undergo over the course of their lives, they have a sense of living exactly one life from the inside.[1] There are several philosophical questions to be asked about this sense of self.[2] One important question is whether anyone *has* a self corresponding to his or her sense of self. That is, we might want to make explicit all the claims about the nature of the self that are implicit in a sense of self and then inquire whether those claims are true of anything.[3] A host of questions follow. Are there selves who lack a sense of self? How damaged can a self be and still be a self? Whatever the nature of the self, what is responsible for that nature? What is responsible for the sense of self (whether or not it corresponds to the real nature of the self)? What might damage that sense? What (and where) are the connections and lines of division between different selves? Must those connections and divisions lie where selves think they do?

These questions are worth asking in part because so much of moral and practical importance can ride on the answers they receive.[4] For example, one is punished for one's own wrong acts, but some types of insanity or ignorance are taken (in law, in conventional morality) to render certain acts not one's own. (We might say they were merely acts of one's body.) This seems intuitively correct, but making explicit why it is correct can be surprisingly difficult. Philosophers interested in personal identity attempt to provide explicit accounts of these and other views of the nature and persistence of the self, both because these accounts are worth having for their own sakes and because they might provide rules that would help us make difficult practical decisions (about punishment, about when someone's past request can cease to govern our behavior). There is a familiar type of nonvicious circular reasoning in play here: philosophers look to the attitudes and implicit views evident in people's confrontations with questions of selfhood and identity in everyday life to formulate their general and explicit theories, and then elements of those theories may be appealed to when making a decision. The hope is that philosophical theory can be both appropriately sensitive to ordinary human realities and of some help in facing difficult realities—that it will at least help us ask better questions if not provide us with firm answers to all the questions with which we start out.

Some philosophical work on personal identity is clear in its aim to affect how we act and how we think (even if it only aims to persuade people to replace whatever conception of the self they once had with the one the philosopher is recommending). But much of the philosophical debate about personal identity takes place at several removes from any type of conversation likely to change action or strategies of deliberation. This is at least in part because there has been and continues to be so much disagreement at what one might think of as the first step in this philosophical work: the attempts to untangle ordinary views of personal identity, present explicit philosophical accounts of personal identity, and determine which, if any, of these we ought to believe. Despite the disagreements, contemporary debates about personal identity take place against a largely common background.[5] Particular philosophical positions can be misunderstood or can seem more puzzling than they are, if their relation to this background is unclear. Two key questions in this background are: How should we understand the logical notion of identity? And how should we understand criteria of personal identity? I will discuss each in turn. My discussion of criteria will involve both epistemological (How do we know?) and ontological (What makes this so?) questions. I will close with a general discussion of the relation between our pretheoretical intuitions about persons and candidate theories of personal identity.

The Logic of Identity

The general notion of identity that philosophers wish to apply to persons is distinct from qualitative identity. Consider a commercially branded object, say, a can of Coca-Cola Classic.[6] I can change some of its properties and still correctly claim to have the same can: I can alter its volume, by crushing it, or its physical location, by moving it, but it is still one and the same can with which I started. Here, *same* is being used to express the notion of *numeric identity*. But if I buy two cans of Coca-Cola Classic, each correctly produced to factory specifications, then I may use the word *same* to express the fact that the cans are alike in their properties—they are *qualitatively* identical.[7] Obviously they are not numerically identical; I bought and paid for *two* cans.

Objects can change their properties—a once shiny penny may become dull; a wrinkled shirt can be ironed. But objects cannot change their numeric identities; the concepts of *change* and *numeric identity* don't fit together properly. We do, colloquially, speak as if we encounter such changes. We say, for example, "She became a different person." Or, "It became a different house." But a philosopher interested in the logic of identity would say that these ways of talking cannot be interpreted as true, literal statements about numeric identity. They might be ways of expressing drastic changes in qualitative identity, such that "She became incredibly vicious" or "It no longer looks haunted" would be acceptable substitutes.[8] Or they might be ways of stating that one item has disappeared and another item of the same kind is occupying the relevant place. If this is what is meant, then the *became different* locution is either misleading or indicative of a special kind of puzzlement. The locution is misleading in mundane cases of ordinary objects being switched. If my cat swaps the ball of yarn I was using for knitting with the one he had previously stowed under the couch, "It's become a different ball of yarn!" is a misleading way to put the point, precisely because it is so obvious that a numerically distinct object is on the scene. (We also have no trouble with the question of where the first ball went when it disappeared—it is now under the couch.)

But suppose we are (apparently) confronted with the disappearance and (re)appearance of persons. In fictional contexts, we allow the possibility of a person dying and another person (or, say, an evil demon) occupying what we once thought of as *her* body. While the body persists, and while we might say, "Its occupant has changed," we feel confident that this is a *substitution* of one person for another, numerically distinct, person—not a change of one person's numeric identity. If someone uses the phrase *became a different person* outside fictional conceits (to describe what has happened to his spouse, for example), we usually expect him to accept *underwent a drastic qualitative change* as an alternative way to express what he meant. If he insists that he didn't mean simply to express qualitative change, unpacking what he does mean will be difficult. In cases when someone's intimates feel that *she* has died though her body remains, regardless of whether or not there seems to be some personality animating that body, there wouldn't have been the wholesale and immediate substitution of persons that is imagined in fairy tales. Even if someone were to insist that his wife died and a different person now lives in the body formerly hers—say, as a result of brain disease or injury—he would still presumably be able to point to specific moments (during the initial progress of the disease or during the coma that first followed the injury) when he wasn't sure who was present. Perhaps this gradualism is part of any appeal of the *became a different person* locution.

It is important to be clear that the reason numeric identity does not (cannot) change is precisely that some qualitative changes to an object can cause that object to cease to exist. While many qualitative changes to an object are compatible with the continued existence of that object *as* that object, not all are so compatible. Some changes are literally destructive. (A hot Coke can is still a Coke can, but a completely incinerated Coke can is no more.) This obvious point can cause problems in cases in which, we might say, two objects occupy the same space—where we have both a *portrait* and a *square of canvas*, say; or, more controversially, both a *person* and a *human body.*[9] When a qualitative change destroys one but not both of the colocated objects—as acid might destroy the portrait but not the (its) canvas, or dementia might destroy the person but not the (her) body—the fact that one of the two objects remains (canvas, body) can make it hard to know what to say about the other object (portrait, person). But assuming the other object was destroyed, the most accurate statement would be that it was qualitatively changed and, *as a result,* it ceased to exist. To say that *it* changed its (numeric) identity would be misleading.

Whatever objects we are discussing—persons or rocks—the relation of identity is defined by certain logical features.[10] Identity is *reflexive* (every object is identical with itself); *symmetrical* (if A is identical to B, B is identical to A); and *transitive* (if A is identical to B, and B is identical to C, then A is identical to C). This distinguishes it from relations such as "smaller than," which is neither reflexive nor symmetrical, and "genetically related to," which, though symmetrical, is not transitive. What makes identity particularly difficult to handle is that, while it is expressed as a relation (e.g., "A is identical to B" or "$x = y$"), the whole point of true identity claims is that there is only one object and so the very idea of *relation* gets only an attenuated grip. (This point is sometimes made by saying that the relation of identity is logically *one-one*.) The terms on either side of the identity sign, or the words that fill in the blanks in "___ is identical to ___," refer to the same object, if the resultant identity claim is true.[11] For this reason, identity claims are usually *interesting* only if the two expressions for the single object come from different contexts (in some sense).[12] Thus, it would be interesting to discover that my boss, whom I've known for decades as "Ms. Smith," *is* my birth mother for whom I've been searching (under the name "Mrs. Jerome Jones"). Identity claims about people often involve this sort of context shift. They also often involve temporal shifts—as when I discover that the girl with whom I once played mud pies *is* my current colleague. The "problem" of personal identity stems, at least in part, from the wide variety of circumstances that render identity claims about people (of this contextual or temporal kind) true and from the variety of justifications for our belief in such claims.[13]

There is a prima facie reason for thinking that answers we give to questions about personal identity will need to preserve the logical features of identity (reflexivity, symmetry, transitivity, and being logically one-one). If we ultimately settle on philosophical theories that do not preserve those features, we need to be clear about what we are doing and why.[14]

Although there is only one relation of identity, such that every object has it to itself, there are many different ways of being an object. It is therefore possible that the details of answers to questions about identity—How many *x*'s do we have? Is this *x* the same as that one? Can you pick out the same *x* you picked out yesterday?—will vary depending on the kind of object at issue. Consider the characteristics of stones, persons, sandwiches, countries, and books. Consider the differences in what is required to produce, maintain, and destroy any of these. Consider the ways we individuate them and count multiples: How many stones are in this box? How many countries are in the European Union? It seems these also will vary with kind. So, what determines how many persons we have (an ontological question)? And how do we know (an epistemological question)?

It may be that the answers to these two questions depend on context. The appropriate notion of "a single person" may be different in legal, moral, clinical, and epidemiological contexts. It may also be that methods of person-counting and person-detecting are appropriately less exact in some contexts than in others. Even so, we might wonder whether there is a single dominant concept of *person* and, if not, whether we need to develop one to fall back on in difficult cases.

As philosophers consider these questions, they also need to attend to whether or not there are minimal criteria for being a person. We don't usually treat all live human bodies as *persons* in any morally loaded sense. That is, not all live human bodies are held fully responsible (either morally or legally) for their actions; not all are permitted to bind themselves through contracts or permitted to vote.[15] Clinical contexts provide many examples of individuals who fall short of personhood in this sense (e.g., they are no longer considered competent to consent to medical procedures).[16] If we employ this morally loaded notion of personhood in asking, in some clinical contexts, whether we have the *same* person as before the onset of disease, the question appears substantive and pressing. For example, in dealing with an individual who has steroid-induced psychosis, we may think it clear that we are faced with *some* person; the question is whether it is the same person, on and off the steroids. Contrast this with an individual whose dementia doesn't cause profound qualitative changes until it also seriously undermines her competence. If we apply only a morally loaded notion of personhood to this individual, the fact that she is losing some of what is required to count as a person (on that notion) will swamp discussion of change. We might conclude that only if one can be counted as a person after one has (for example) lost the *legal* status of a competent person will some clinical cases seem to pose the question: Are we faced with the *same* person as before the onset of disease?

Complications

I have discussed the constraints imposed by the logic of identity on attempts to answer epistemological and ontological questions about the entities we count as persons. Before discussing the content of the answers the philosophers explore, I want to examine two potential challenges to progress in this area.

The first is posed by what is called the *paradox of analysis*. When one attempts to provide an analysis of a concept, there is a question about whether one can be both informative and accurate. To be informative, the analysis needs to provide something that was not (at least not explicitly) present in our pretheoretical, prereflective use of the concept. And to be accurate, the analysis needs to be an analysis of the concept in question—one can't, for example, introduce a new (or technical) concept, analyze it, and claim to have provided an analysis of a concept in common use.[17] But it isn't obvious that every concept in which we might be interested can be analyzed both informatively and accurately. Some philosophers hold that nothing other than the concepts *person* and *personal identity* can be displayed, in an analysis, to be doing the work done by those concepts.[18] We can, of course, talk about them, give examples of good and poor applications of them, and trace their connections with other concepts. But that's all. What we cannot do is find a more tractable substitute for either of these concepts, such that (for example) *has the same brain as* turns out to be a perfectly good substitute for *is the same person as*.[19] Those philosophers who think an accurate and informative analysis of the concept of personal identity is possible face a question about whether such an analysis should be construed reductively. For example, should we say, after analysis, not only "Personal identity is constituted by psychological continuity," but also "Personal identity is *nothing but/beyond* psychological continuity"? The test for those who want to say yes to that question is: Could we describe the world, leaving nothing out, without using the concept of personal identity and using only (in this example) that of psychological continuity?

The second challenge involves the costs and benefits of revising common sense about such a key notion as personal identity. Individuals and cultures have some implicit sense about what is required for being a person and what is required for the truth of identity claims such as, "This is the same person as that one." It is possible that philosophical investigation of questions about personal identity may reveal contradictions or tensions in these implicit notions. It is also possible that philosophers may propose notions of personal identity that are free of those tensions. The question of when common sense should be revised and when, rather, the divergence of common sense from philosophical theory should be taken as a sign that more philosophical work needs to be done is not easy to answer. Perhaps the apparent contradiction revealed by philosophical reflection is only apparent; or perhaps it is a sign of genuine, irreducible paradox, such that it is a mistake to embrace a tension-free account. Even if revision is undertaken, it isn't clear how far (into which contexts) it should be taken. What should happen, for example, if a clinician or hospital ethicist with a "revised" notion of personhood and personal identity is counseling a patient or family member with an unrevised notion? Whose notion has priority, and are there any but pragmatic reasons for deciding such a conflict one way or the other?[20]

Both the paradox of analysis and the question of revisionism raise questions about the status of our pretheoretical intuitions about personal identity in normal cases. Can these intuitions serve as guides to difficult cases and abnormal circumstances? That is, can reflection on them help us choose a particular theory of personal identity that can then be applied to the difficult cases? One might answer no, if one thinks that such intuitions are simply the emotional or practical expressions of a theory that we already (implicitly) hold and that can't easily be applied or extended beyond normal cases. The trouble is that abnormal, difficult cases do arise, and we want some kind of guidance for dealing with them; it isn't clear where we would start if not from our intuitions about the nature of persons and identity as revealed in ordinary circumstances.

The Importance of Identity

We have several pretheoretical assumptions about what is at stake in the question of personal identity. Philosophers rely on these assumptions in making their arguments. They expect us to have certain responses to the thought experiments they propose, for example, or to find certain types of revisionism more acceptable than others because of how relatively little the recommended revisions disturb these assumptions. In the tradition from which all three philosophers contributing to this volume draw, there is general agreement that a limited number of factors explain why we care about identity *as* we do. (It is a separate question whether we are correct to care.)

These factors cluster into three interrelated sets of phenomena and concerns. The first set concerns *survival* and our desire for our loved ones and ourselves to survive.[21] Persons suffer death, and in ways other than by the absolute destruction of the human body. So for any proposed change, one can ask of the person about to undergo it, "Will he survive this event?" Most people could probably agree on a list of changes (e.g., religious conversion, amputation of an extremity) that persons clearly survive and a list of changes (e.g., destruction of the heart and brain) that persons clearly do not survive. The difficult questions about what counts as survival emerge in cases such as a sudden descent into profound mental illness or a lapse into a persistent vegetative state. Even family members can disagree about whether their loved one has survived a change of this type.[22]

The second set involves *anticipation* or *self-interested concern.* We worry about and act prudentially on behalf of many people. But the worries one has and the precautions one takes for oneself seem to differ in kind (even if not in urgency, moral importance, or emotional salience) from the worries one has and precautions one takes for others. For which future events should one have self-interested concern, and for which other-directed concern? Which events are events of one's own future? (For example, my daughter's infant immunizations pain me, and her being vaccinated is an event in my future; but I am not vaccinated by her pediatrician. *Being vaccinated* is an event in her immediate future, not mine.)

The third set involves *responsibility* and *desert.* The very idea of action (as opposed to mere behavior) seems to involve responsibility, if not intention. Thus, it is a trivial truth that each person is responsible for his or her own actions. One deserves praise and blame, punishment and reward, for acts for which one is responsible. But how does responsibility carry across time or significant qualitative change? What makes it appropriate (if it is appropriate) to punish someone now for acts committed in 1977 or while he was intoxicated? We assume that wages may be paid monthly rather than (say) hourly because we assume the one working *is* the one getting paid; but can we justify that assumption?[23]

Survival, anticipation, and responsibility seem to depend on identity and seem to reveal what makes questions about identity so morally and practically important, not to mention emotionally uncomfortable. Reflecting on the circumstances in which we say someone didn't survive some event, or someone was wrong to fail to anticipate some pain, or someone ought to admit responsibility for some act (and so on) should help reveal the view of personal identity with which we are pretheoretically operating. We can then consider whether it ought to be improved.

Criteria Coming Apart

We have, ordinarily, many criteria to offer in response to both ontological questions (e.g., What makes a person the same person she was 10 years ago?) and epistemological questions (e.g., How can we tell if this is the same person we treated here 10 years ago?). The concept *criterion* is appropriately used in pursuing either ontological or epistemological questions, although something could be evidence for identity without constituting identity. Ordinarily, one won't have good evidence for two persons being the same if they in fact are not—but not every case is ordinary. Also, ordinarily, in answering either type of question, relying on any one of the relevant criteria is just as good as relying on any other. That is, you will arrive at the same answer regardless of which criterion you choose, and each criterion is about as intuitively appealing as any relevant other. For example, suppose some clinicians have a question as to whether a given patient is the same woman they treated two years ago. A comparison of fingerprints, the ability of a nurse to recognize her, and the patient's obviously accurate memories of what the hospital was like back then would *all* prompt a yes to the question of whether this is the same woman, and (barring amnesia and fibbing and so on) all look like good reasons for answering yes.

But of course, the ordinary isn't all we have to confront. Criteria don't always lead us to the same initial answer when we apply them to a difficult case. (For example, an amnesiac patient might be unmoved by fingerprint evidence that she had been hospitalized before.) Some philosophical work on personal identity seems to be of limited practical use, because it relies so heavily on fantastic thought experiments. But the point of the experiments is simply to probe our intuitions about difficult cases, particularly cases in which criteria—in both the evidential and the constitutive sense—for identity point in different directions. When thought experiments are well designed, they press exactly where we ought to be pressing: at the intersection between empirical questions about what we can discover and conceptual questions about what we (in some sense) decide.[24] The case studies presented to the contributing philosophers were intended to provide this sort of pressure.

Roughly, criteria for answering (epistemological or ontological) questions about personal identity sort into two categories: the physical (those more or less concerned with the states and changes of live human bodies) and the psychological (those more less concerned with the states and changes of minds, or "psyches" or "consciousnesses").[25] This is true both of common sense (broadly speaking) and of philosophy. Philosophers form a "view" of personal identity by selecting the criterion or criteria they take to be fundamental to it. So the views divide roughly into the psychological and the physical.[26]

Particular bodily criteria can fail to track personal identity. Some narrow method of reidentification, such as fingerprinting, can be circumvented (fingertips can be altered or destroyed) and yet the person remain. More fantastically, consider the "genetic therapy" undergone by the evil genius in the James Bond film *Die Another Day,* who appears to change his ethnicity. Standard physical criteria for identity would all suggest that Bond is correct in his initial assumption that his enemy is no longer in the vicinity, but we in the audience know that, *really,* the evil criminal is still here, that Bond must outwit the *same person.* Conversely, bodily stability can coexist with psychological change, including the dramatic changes caused by brain injury and disease. In these cases, standard physical criteria of identity suggest we have the same person over time. But the psychological changes can be so profound that, especially if we implicitly subscribe to a psychological view of identity that emphasizes memory and character, we may wonder whether it is legitimate to think we are faced with the death of one person followed by the arrival of another.

Sometimes when criteria of personal identity come apart, it is *obvious* that one criterion (or small set of criteria) should "trump" the others.[27] This is certainly what a viewer of *Die Another Day* is supposed to feel: psychological continuity trumps all the physical differences between (what we therefore refer to as) the villain's old and new bodies.[28] In other cases, at least at first glance, we simply seem to have criteria competing with no resolution in sight. Suppose a woman with frontotemporal dementia has become violent and reckless. Her husband feels that his wife is (in some sense) *gone.* He wonders what trying to be a loving spouse amounts to in this case. Of course, he knows that by most physical and even many psychological criteria of identity, the woman living with him is the woman he married 40 years previously. But if he concludes he has the same duties to her as to the woman he married, this will be the result of painful reflection, not any obvious inference from similarity of fingerprints. There just isn't a single fact one can appeal to that dissolves all difficulty. The relative weight of the competing criteria isn't immediately clear.[29]

The possibility of competing criteria of personal identity raises many questions. Across contexts, or in some given context, is there some criterion that *objectively* trumps all others? Is there an "essence" to personal identity (whatever sort of thing it is that fulfills the criteria for being the same person)? How do we find out? If not, what then? Are there better and worse criteria for each context? Are the criteria relative to the interests operative in a given context?

Are they relative to the persons involved? (For example, should a clinician help the spouse of a woman with dementia explore which criteria of personal identity were implicitly important in the context of the life he lived with her before she became ill? And should those criteria guide medical decisions concerning her?)

One way to begin answering these questions is to test our intuitions about various scenarios, both real and imagined. When philosophers propose such scenarios for consideration, what they are usually seeking is a clear intuition about the importance of identity that can then be used to argue that one criterion of identity is clearly more fundamental than others. For example, someone might cite the willingness of movie audiences to hold the visually Nordic (posttherapy) individual *responsible* for the crimes of the visually East Asian (pretherapy) individual as evidence that psychological criteria trump physical ones (at least in the theories of personal identity implicitly held by people who enjoy Bond films). Or you might consider the quality of your fear of developing frontotemporal dementia as revealing that you fear not *surviving* it. You might then conclude that you implicitly hold a theory of personal identity in which persons can die in more ways than persons' bodies can.

Physical Criteria

If we adopt a view in which physical criteria are constitutive of personal identity, then whenever we allow that we are faced with the same human animal (same live human body), we will conclude that we are faced with the same person.[30] Adopting such a view makes it easy to deal with the central questions in basic cases, in part because it allows us to rely on physical criteria so as to generate clear, unambiguous tests for identity (e.g., in matching fingerprints or dental records). But none of the philosophers contributing to this volume would accept *human animal* as an analysis of *person*, nor would they accept that personal identity reduces to (human) animal identity.[31] In fact, most contemporary philosophers reject the idea that any nonpsychological criterion could constitute personal identity. Human persons are embodied, however, and paying adequate attention to this fact while holding a psychological view of personal identity can be challenging.[32] If persons are not identical with live human animals, how do we classify features of persons and features of animals? Do we have *two* consciousnesses in the same location of my physical body, one "personal" and one "animal"? And how do we talk about human development and decline? What is my relation to the (animal-but-not-person) embryo my mother carried before I was born? Was *I* once that embryo? If yes, how did I achieve personhood? If no, where did I come from?[33]

These are difficult questions, and anyone endorsing a psychological view of personal identity must eventually deal with them. In my view, we already—in ordinary life—implicitly treat physical criteria as the chief evidence for identity, but treat (at least some) psychological criteria as constitutive of identity. Certainly, psychological criteria appear to be more deeply connected than physical criteria to the issues of survival, self-concern, and moral responsibility.

Psychological Criteria

The human psyche is complex, and not all of its aspects, features, and capacities will be equally important to personal identity. Philosophers developing psychological views of personal identity make judgments about which of the many psychological criteria that are relevant (at least in an evidential sense) to personal identity might actually be constitutive of identity. (Care is required because the criteria that are easiest to use in establishing identity in, for example, simple imposter cases are not likely to be central enough to be constitutive: I may immediately begin testing someone claiming to be my long-lost cousin about details of our shared childhood, but she could be my cousin and fail to remember; and she might sincerely take herself to remember and not be my cousin.) Psychological views of personal identity frequently appeal to the notion of psychological continuity: to the links of memory, emotion, and character development that hold within a self.[34] The trouble with taking psychological continuity to *constitute* personal identity is that it lacks the logical features of identity. Some forms of it are not reflexive (I remember my 5-year-old self, but she doesn't remember me) and it can't easily be made transitive (an 80-year-old may remember the feelings of her self at 50, and her self at 50 may remember her feelings at 20, but her self at 80 may have no recollection of that 20-year-old's feelings). Most important, it is not logically one-one: a single person could be psychologically continuous with two or more persons. This point has been made by several thought experiments.[35] I'll run through a schematic experiment and then discuss how reactions to it shaped the development of different philosophical views.

Choose whatever physical condition you like that you are halfway willing to think supports your psychology (you could choose your brain, or anything with the same connectionist architecture as your brain, or whatever else you find plausible).[36] Then suppose there are technical methods for getting an individual to be in, or to have, the physical condition that you are now in or have and that supports your psychology.[37] That means your psychology could be produced in two (or more) physical locations. The thought experiments imagine us using either a low-tech or a high-tech means to do so. In the low-tech version, we assume that each hemisphere of your brain could independently support your personality; we then imagine that a surgeon transplants each hemisphere into a human body and destroys "your" body (the one whose skull originally housed the now divided brain). In the high-tech version, we imagine it is possible to record perfectly the structure of your brain and then to place matter with that structure inside the skulls of two human bodies and destroy "your" body.[38] We then ask: Where are *you* (if you exist at all)? The final part of the experiment requires you to imagine that one of the transfers or productions was unsuccessful (a clumsy surgeon drops the second hemisphere; a clumsy computer programmer permanently erases essential code after only one of the new brains is produced). We then ask again: Where are *you* (if you exist at all)?

When I contemplate any of this happening to me, even if I am told to assume that the resultant individuals will act and think like me and will feel themselves to be psychologically continuous with me, will *say* that they *are* me, I just feel sad. I don't, in my gut, think I survive any of this. But many philosophers have claimed (more or less) that most people have (or should have) the following intuition: if the physical condition that supports your psychology is produced in *one* human body, you survived; if it is produced in more than one, you don't. (How can there be two of *you*? How can you be identical with each survivor when they are obviously not identical with each other?)

And the real philosophical work begins at this point: with reflection on why the intuitions would (at least initially, for many people) differ, depending on the presence or absence of duplicates. Several very different views began from the reflection that it seems odd for personal identity—something that is supposed to matter so much—to depend on such extrinsic facts as whether a clumsy surgeon dropped one brain hemisphere on the floor, thus preventing a "double."[39] One can simply accept that personal identity is fragile in this way and continue to take psychological continuity to constitute identity; this necessitates including a "no duplicates currently exist" clause in any analysis one provides about what makes person B identical to person A.[40] Or one can say that any phenomenon this fragile can't be what we were aiming to discuss when we started off with our serious questions about survival and moral responsibility: so either this phenomenon isn't personal identity, or it is, but what really matters is something else. Another possibility is to suggest that we mistakenly overvalue identity and that for all moral and practical purposes, including prudential self-concern, what matters is psychological continuity.[41] Whatever view one endorses, one needs to defend it against philosophical competitors and give some account of its relation to our extraphilosophical intuitions and attitudes. As news of developments in psychology, psychiatry, and neuroscience is increasingly infiltrating even "commonsense" views of the self, it is important for philosophers to place their theories in relation to those disciplines and their effects on ordinary attitudes.

Conclusion

It is a commonplace that observation can change the observed. If the object of observation and theoretical speculation is the sense of self, such changes seem likely, especially if active narration is a key component of personal identity.[42] Taking up new theories of the human self complicates the task of integrating our sense of self with our awareness of the world beyond ourselves, including others' sense of us. A question posed by the case studies is: How ought we to view changes in ourselves when those changes are not easily construable as the results of our own intentional activity? How ought we to view such changes in our loved ones?[43] What is involved in anyone's sense that a personality trait, family commitment, or career goal is *hers*? In particular, how can, how should, knowledge of the ways in which neural structures support one's sense of ownership affect that sense? These sorts of questions may shape anyone's task of self-understanding, though they will seem especially pressing when a self seems under threat.

When someone's competence, capacity for autonomy, and personality are damaged or altered by disease or injury, his or her intimates and physicians face a host of difficult decisions. If philosophical theories are a help to people faced with such decisions, it isn't likely to be because they are explicitly drawn on in moments of crisis, and it certainly won't be because they provide answers that obviate the very need for decision. Rather, they may help prepare the ground for deliberation and provide a framework for asking questions and keeping track of competing concerns. Perry, Rovane, and Schechtman each suggest (for different reasons) that being a self is a job of work; and each offers a way to frame detailed questions about how to respond to those people who are losing or have lost the capacity to do that work. Asking those questions with care is itself a kind of compassionate treatment.

ACKNOWLEDGMENTS

This chapter began as a presentation at a seminar meeting of the Program in Ethics and Brain Sciences in September 2004. I benefited from the questions and comments of seminar participants. Debra Mathews and Patrick Duggan provided helpful comments on a later draft.

NOTES

1. John Locke (1975) claimed that "every intelligent Being, sensible of Happiness or Misery, must grant, that there is something that is *himself*, that he is concerned for, and would have happy."

2. There are also some psychological, psychiatric, and neurological questions to be asked, of course (e.g., When do children develop a stable sense of self? What causes personality disorders? Which regions of the brain are active during self-referential thinking?). And sometimes a single question (e.g., What can damage the sense of self?) can receive distinct, not incompatible, answers from different disciplines.

3. For example, many people would describe themselves as having a stable, substantial self whose changes they can observe and report. This suggests that such a self exists prior to and in some sense independent of these activities of self-observation and self-description. But the contrary suggestion, that the idea of such a self and the intimate sense of its presence are both artifacts of these activities, has a long philosophical pedigree. For a prominent defense of this view, see Section VI, Book I, of David Hume's *A Treatise of Human Nature* (1978).

4. This point will be complicated below (see the second paragraph of n. 41).

5. All three philosophers contributing to this volume belong to the tradition of English-language analytical philosophy. I will use *philosophy* and its cognates to refer to work within this tradition, broadly construed. This is not intended as a dismissal of other philosophical traditions.

6. The usefulness of branded objects to discuss the distinction between numeric and qualitative identity occurred to me as a result of an example of Hilary Bok's.

7. The cans are alike in all but the properties that are connected to their being numerically distinct. For example, if I hold one can in each hand, then one can will have the property of being held in my left hand, and the other will lack that property at that time. These sorts of differences are not usually taken to undermine qualitative identity.

8. In both these cases, it seems important that we have a general familiarity with the possible causes of such changes, even if we don't know what caused the transformation in some particular case. Houses are renovated, for example, and people's characters are sometimes dramatically altered by tragedy.

9. It is relatively uncontroversial to distinguish between the *concepts* of "human animal" (or "live human body") and "person." As soon as one claims that the concepts do not have the same extension—that is, that the concepts do not pick out or apply to the same objects or classes of objects—one has made a substantive and potentially controversial claim.

10. The following discussion assumes that there is only one relation of identity. This is the dominant but not the only philosophical position on the subject. For an argument that there are kinds of identity (and not just kinds of objects all related to themselves by a single identity relation), see Geach 1980 [1967]. The discussion also assumes that it is worth asking how the logical relation of identity applies to persons, if only because that question dominated the debate for such a long time. All three philosopher-contributors to this volume would agree that even full answers to questions about logical identity of persons won't answer all the serious questions we have about persons and personhood. Of the three, Perry is perhaps most comfortable applying the logical notion of identity to persons, in part because he has a deflationary view of its importance. Schechtman is more emphatic that the notion of logical identity, applied to persons, is not going to help us answer questions about the creation and maintenance of selves. While Rovane doesn't explicitly disavow the usefulness of logical identity, its usefulness to her account is attenuated because of her view that personhood comes in degrees. (She doesn't think the question "Is A the same person as B?" always has a simple yes-or-no answer.)

11. That is, in "$(3 + 5) = (2 + 6)$," "$3 + 5$" and "$2 + 6$" both pick out the number 8; and in "Norma Jeane Baker is Marilyn Monroe," both "Norma Jeane Baker" and "Marilyn Monroe" are names of the same woman.

12. Gottlob Frege (1997 (1892]) is usually credited with developing an account of this in his paper "On *Sinn* [sense] and *Bedeutung* [reference]."

13. Consider how relatively few kinds of circumstance are relevant to identity claims about balls of yarn, and how relatively obvious are the causal mechanisms involved in those circumstances (cats batting them across rooms, children dropping them into the fireplace). For a discussion of the relation between our understanding of causality and questions about identity, see John Perry's contribution (chapter 6, appendix).

14. How does it even make sense to suggest that an account of personal identity could fail to preserve the logical features of identity? It makes sense if to *give an account of personal identity* is to answer, or provide the means to answer, the collection of particular questions that we have about persons, their natures, and their persistence conditions. It may be that answering questions about dementia, mental illness, and profound character change, and all the other puzzling phenomena generated and suffered by persons, will not in the end require us to rely much on the logical notion

of identity. In that case, we might describe ourselves as wanting some answers about persons, assuming these answers were to be found by applying the notion of identity to persons, and discovering that this wasn't so. Confusion would arise only because we used *personal identity* to refer to the phenomena for which we wanted a satisfying account (of whatever kind) *and* to refer to the identity relation when persons are the objects named by terms used to fill in the blanks in claims of the form "___ is identical to ___."

15. Someone who fails to count as a person in this sense will not be allowed to take up certain responsibilities (to society, to other individuals, to herself). The questions of what is morally owed to this individual and what responsibilities others have to her are distinct. For discussions of what is implied by conferring the label of *person* on an individual or by withdrawing it from her, see the contributions by Marya Schechtman (chapter 4) and Carol Rovane (chapter 5).

16. See Samuel Barondes's discussion (chapter 7) of the Global Assessment of Functioning (GAF) used by psychiatrists and psychologists to place individuals on a continuum of functioning.

17. Of course, explicitly recommending that we operate with some concept *C* rather than a related but distinct concept CPR is perfectly acceptable. What is not acceptable is to represent that recommendation as an *analysis* of PR. (It might be acceptable to represent it as a way to do better what we want [or ought] to do with CPR.)

18. See, for example, Chisholm 1976.

19. *If* brains are taken to be individuated in something like the obvious way that rocks are, then "same brain" *is* easier to understand than "same person." Thus, if it turned out that this were an *accurate* analysis, it would be informative in that (1) we would be able to give a fairly clear account of personal identity, and (2) we would have acquired a useful strategy for dealing with difficult cases. (Just open up the skull and compare what one sees to, say, a picture taken in the past.) But of course, anyone seriously proposing "same brain" as an analysis of "same person" would not be thinking of brains as being like rocks in this way. And once brains are considered as dynamic, functional entities, it isn't clear that answering ontological or epistemological questions about their identity will be simple. So the tractability gain would largely disappear. Perry, for example, takes our questions about persons to be questions about minds and believes that the mind is the brain. Rovane, by contrast, does not think that an account of personhood is an account of the activity of the brain (though, of course, human persons cannot exist without functioning brains). And yet it isn't the case that Perry's theories enable him to make short work of the problems posed by the case studies. If one prefers Perry's account, it won't be because it makes everything simple.

20. Rovane's project is explicitly revisionist. She thinks revisionism can't be helped, because the "common sense" on these matters contains serious contradictions. Schechtman doesn't want to be a revisionist; she thinks it is better to throw out theories that conflict with intuitions. Perry is ambiguously revisionist; he takes himself to be respecting and accounting for the roles we pretheoretically ask identity to

play, but he does recommend some revisionism in our view of the connection between identity and those roles.

21. I use *survive* and its cognates to refer to earthly survival—the continuation of the type of life that, even for someone with religious faith in life after death, is assumed to end for every human person at some point.

22. Rovane explicitly recommends that some clinical procedures be described to a prospective patient as involving the death of her self, followed by the birth of a different person in what is now her body. For a related discussion, see Perry (chapter 6).

23. Almost any contractual or quasi-contractual issue will have the assumption of the identity of the relevant parties at its heart. For example, if I sign an advance directive today, it is binding on clinicians 20 years from now because (it is usually assumed) I am the same person, however impaired, at the later date. To highlight the assumption, consider a different rationale for holding the clinicians to the terms of the directive: "the wishes of the person I am now ought to govern the treatment of any human body continuous with the body that embodies the person I am now." This doesn't make any claim about identity of persons, not even if we add "because there is no one better placed to dictate this."

24. For a more skeptical view about the usefulness of thought experiments, see Rovane (chapter 5). How to balance empirical and conceptual aspects of personal identity is a problem not only in abstract philosophical theorizing but also in clinical and other contexts for practical decision-making. For example, one can do empirical tests of short- and long-term memory. Concluding that a patient has deficits severe enough to prevent her from maintaining a persistent sense of self, however, would involve not only attention to empirical data but also a thesis about which cognitive functions are essential to the sense of self and which could be replaced or managed without. And the formation of theses of this kind requires (at least implicit) decisions about what is essential and what peripheral to the normal sense of self.

25. This division into psychological and physical is intended to capture some distinction that everyone, whatever his or her ultimate view about (for example) the relation between the brain and the mind, will recognize as important. That is, it is not supposed to imply that something *other* than the biological body is causally responsible for mental life and human behavior. It is simply a rough and ready way of getting at the distinction between questions about fingerprints or facial contours and questions about memories or temperament.

26. I haven't mentioned the soul. Western philosophers not working in a religious tradition (whatever their own beliefs) have tended to accept that "identity of soul" is not a useful analysis of personal identity. (Locke, for example, argued that what really matters is continuity of consciousness, not sameness of body or of immaterial spiritual substance [Locke 1975, bk. II, chap. 27, §§ 23–26].) This may be because they think there is no such thing as an immaterial soul, linked to the body and distinct from the mind, that the body (somehow) causally supports. But it could also be because they think that talk of soul-criteria is effective, if it is, by piggy-backing on psychological criteria that then don't get properly analyzed. Pressing on the psychological

criteria would then be a way of continuing the analysis of the concept of identity, *not* a commitment to a "no-soul" view of persons. In my view, an emphasis on psychological criteria of personal identity—unless, perhaps, one were aiming for a truly reductive analysis of identity—is neutral on this and other theological matters.

27. One might say that deciding on a diagnosis of amnesia is tantamount to deciding to treat physical criteria and other people's memories as obviously trumping any criteria related to the patient's current psychological capacities.

28. Of course, the viewer (unlike Bond) is able to observe some of the medical procedures that result in the dramatic change in the villain's physical appearance. What happens to him is, really, just thoroughgoing plastic surgery—down to the cellular level. So our willingness to think there is psychological continuity may depend, in part, on the fact that we are able to imagine his body undergoing gradual qualitative change. Even though the beginning and end results are dramatically different—such that someone relying exclusively on physical criteria of identity would *appear* to have good reason to think these were two different people—there is an explanation that takes us from one to the other. Does this mean that we take psychological continuity seriously (as possibly constituting identity) only when some kind of physical continuity, however attenuated, is preserved? I'm inclined to think not. It is true that when psychological continuity is not linked with physical continuity, as it normally is, we do want some explanation of how this is possible. But we want the explanation for its own sake, not to support the very conceivability of psychological continuity.

29. I am making a point about the structure and complexity of the metaphysical and moral questions in such cases. I am not suggesting that such questions never have correct answers, nor am I dismissing the possibility that a spouse might feel some single fact to be obviously determinative. I am, however, committed to the claim that such a spouse has made an implicit decision to treat that fact in that way.

30. See Olson 1997. It is crucial that *human* here should be understood in as strictly a biological sense as possible. That is, while many of the features that differentiate *Homo sapiens* from his near primate ancestors cannot be described without mentioning the activities (e.g., linguistic communication) that *Homo*'s distinct physiology supports, this cannot be taken as license to pack psychological criteria of personal identity into the conditions for being a human animal. Otherwise, these views would cease to be distinctive. In particular, they would become difficult to distinguish from the views of someone like Perry (see n. 31).

31. Schechtman would be sympathetic to granting physical criteria some serious weight. She emphasizes the relation between social roles and personal identity and argues that our embodied natures impose constraints on how far apart one's view of oneself and others' views of one can be. Rovane strongly deemphasizes physical criteria because she doesn't think divisions between persons track divisions between human animals. Her view that personhood is an achievement, a product of decision, also means that she will reject any claim to the effect that some physical structure *suffices* for personhood. Perry plumps for psychological criteria over physical criteria in the

sense that, in his view, it is mentality that matters to personhood and personal identity; but he is ultimately committed to giving a physicalist account of minds, in the sense of claiming that minds are brains. It is important not to collapse these two distinct positions of Perry's.

32. Rovane's theory easily accommodates both the possibility that many persons could share a single human body and the possibility of a "corporate person" with many bodies. It could, in principle, accommodate nonembodied persons (e.g., God). She nevertheless faces the questions that follow in the text here, because she takes it that most persons are, as she puts it, "of human size."

33. These questions will only arise for someone who assumes some relevant notion of personhood in which an embryo lacks personhood for at least some time.

34. Of course, one of the main challenges of such a view is to flesh out exactly what psychological links hold within instead of between selves. While I probably can't feel guilty except about my own actions, I can be shamed by the actions of my friends and family as well as by my own. Is there something distinctive about the shame I feel due to my own actions, apart from its having been caused by my actions? What makes my shame (my passions, my memories) *mine*? If psychological links—between thoughts, feelings, memories, intentions—are supposed to constitute personal identity (and not simply result from it), then we can't merely add self-referential terminology (*my* memory of *my* shame, *my* anticipation of *my* pain) in attempting to explain the distinctive links between psychological states that hold within a self. For some discussion, see Shoemaker 1963, 1970.

35. For some classic presentations, see Nagel 1971; Parfit 1971; Perry 1972. There is no doubt that the scenarios drawn on in these discussions are far-fetched and that philosophers are not always careful to distinguish (1) the claim that a scenario (and the philosophical points suggested by reflection on it) was suggested by some development in neuroscience from either (2) the claim that a scenario, as presented, was drawing on neuroscientific facts or (3) the claim that neuroscience was now in a position to demonstrate the truth of some philosophical thesis. Philosophers ought to be clearer about which, if any, of these claims they intend to put forward. But readers of philosophy need to be careful to respect the distinctiveness of the first claim. For example, suppose a philosopher asks, on reading about Michael Gazzaniga and Roger Sperry's work with split-brain patients, "What does thinking about the possibility of two distinct centers of consciousness in one cranium reveal about our pretheoretical notion of personal identity?" It would be unfair to represent this philosopher as thinking that Gazzaniga and Sperry had *shown* that, in fact, individuals whose corpus callosum had been cut possessed two whole and discrete minds. (For a discussion of recent research on split-brain patients exploring the sense of self, see Gazzaniga 2005.)

36. I use *support* in the sense of "causally enable." This locution does not suggest that the psyche is some mysterious item over and above its causal supports, but nor does it imply that psychology reduces to these causal structures.

37. If you believe the brain supports your psychology and that only by living your

life over again could a brain identical to yours at birth be identical to your brain now, you will refuse to grant this possibility.

38. To give the flavor of an extensive literature, I am deemphasizing some important issues. For example, if one endorses a version of the psychological view of identity that emphasizes the role of the normal functioning of the human body in supporting our psychology, one might want to insist that the hemisphere-transplant case and the computer-assisted teletransport case be treated separately. I'm not discussing this or any other way in which the causes of the relevant psychological continuity might be taken to matter to identity, independent of the continuity they produce (see Parfit 1984, chap. 10). Also, the thought experiments all rely heavily on the assumption that preserving or replicating certain physical structures suffices to preserve or produce the psychological constituents of identity. Rovane would certainly reject that assumption; Schechtman would not endorse any simple version of it; and while Perry would probably endorse it, this isn't to say he would think it needed no defense (see n. 31).

39. Of course, the following views could be argued for without thought experiments. But the split-brain and teletransporter scenarios were prominent in the philosophical literature on personal identity in the last 30 years and were often responsible for convincing individual philosophers of one or another view.

40. A notional variant of this view claims that psychological continuity constitutes personal identity—that is, for person B to be identical with person A *just is* for person B to be psychologically continuous with person A—only if the causes of the psychological continuity are of the right sort. The difficulty, of course, comes in trying to decide what is a cause of the right sort. Aging from 30 to 70 would be a cause of psychological continuity of the right sort. A cause that can generate two psychologically continuous persons from a single person won't be counted as normal, but it can be difficult to make this look anything but question-begging.

41. Here is a rough placement of our three contributing philosophers. Schechtman thinks it isn't really criteria for the logical relation of identity (for persons) that interest us in connection with any serious moral, practical, or metaphysical questions about persons. She takes it that those questions are distinct from the question "When is it correct to reidentify this person as that one?"—for which the logical relation of identity does matter. Perry supports a moderate version of the idea that what we ought to care about is psychological continuity. Because Rovane argues that what matters for personhood and personal identity is rational unity and not psychological continuity, such that single persons can involve more than one human animal (and its psychic life), she can deal with duplicates as well as she deals with group agents. For Rovane, the two resulting individuals will be separate persons if they don't commit to rational unity and a single person if they decide to so commit. The status of the original person with respect to the resultant person(s) would be handled, presumably, analogously to her handling of group agents when the number of human animals involved changes over time (e.g., when members leave a club whose commitment to rational unity gave it the status of a person).

Derek Parfit argued in *Reasons and Persons* (1984) that everything important about personal identity was captured by psychological continuity *and* that this fact could be the foundation of an argument in favor of utilitarianism over ethical views that take the distinctness of persons seriously. This is an example of taking ethical conclusions to follow from independently determinable metaphysical facts about persons; Rovane explicitly objects to this procedure. Schectman never formulates such an explicit objection. However, when she discusses degrees of personhood, the capacity for moral agency, and changes of personality with reference to the case studies, she relies on the moral responses of the patients' wives as evidence for the metaphysical conclusions she draws. Perry, too, takes a version of this from-ethics-to-metaphysics tack in his treatment of the Alzheimer disease case study.

42. See Rabins and Blass (chapter 2).

43. See Perry (chapter 6); Rovane (chapter 5); Schechtman (chapter 4).

CHAPTER TWO

Toward a Neurobiology of Personal Identity

Peter V. Rabins, M.D., M.P.H., and David M. Blass, M.D.

Because mental experiences derive from the brain, it has long seemed likely that neuroscientists could make progress in understanding the neural basis of specific mental experiences. In this chapter, we review three approaches to advancing our knowledge of how and which brain structures and functions might contribute to the experience of individual or personal identity. One approach, neuropsychiatry, is clinical and is derived from "accidents of nature," that is, injuries to and diseases of the brain. The second approach, experimental neuropsychology, stems from the study and manipulation of normal (intact-brain) and brain-injured individuals. The third, developmental psychology, is both descriptive and experimental and primarily uses information gathered from the study of normal infants, although some data from adults and from individuals with impaired function are also cited. Such approaches can illuminate plausible central nervous system underpinnings of an experience such as personal identity, but they are unlikely to explain the construct and experience as a whole. Whether other approaches, in addition to these, can do so is beyond the scope of this chapter, but concepts such as personal identity have an ineffable aspect that will never be fully encompassed in a strictly neurobiological model.

Neuropsychiatry is the discipline that seeks to learn about brain-behavior relationships by studying diseases with identifiable brain pathology that are associated with changes or impairments in mental life. This method has been used to study the neural basis of human cognition and behavior since Broca's demonstration in 1869 that left-hemisphere stroke is associated with the development of a language disorder. An example of its application to a noncognitive sphere is the study of major depression in persons with stroke (Robinson 1998), multiple sclerosis (Rabins et al. 1986), and Parkinson disease (Marsh 2000).

The term *personal identity* is used here to refer to the reports of individual human beings that they experience themselves as unique individuals *and* to others' observations of the unique characteristics of that individual. We distinguish personal identity from *personhood,* which refers to the concept that the individual life has a value vis-à-vis its status as a living being. We note that there is overlap between the constructs of personal identity and *self,* but suggest that *self* refers primarily to the experience of the individual, whereas *personal identity* also includes the views/observations of others.

The question of whether a construct such as personal identify or self-awareness can be localized to the brain reflects a two-century-old argument about whether specific cognitive, emotional, and behavioral capacities are located in identifiable regions of the brain or brain functions are diffusely distributed. There is now evidence that both sides are correct and that the answer differs by construct. For example, language is primarily subserved by neurons located in the left frontal lobe and left temporal lobe in most (approximately 95% of) individuals. Remote or long-term memory, by contrast, seems to be more diffusely stored. Thus, this chapter examines not only whether there is evidence that personal identity is located in the brain but also, if it is, whether there is evidence of focal and/or diffuse structural and functional contributions. In fact, the data support the intermediate position that some aspects of personal identity can be localized but that personal identity as a whole requires the interaction among a distributed set of brain locations and systems.

Neuropsychiatric Evidence

Impairment of Self-perception (Agnosia)

Several abnormalities of self-recognition have been known to clinicians for more than 100 years. In 1891, Sigmund Freud used the term *agnosia* to refer to a person's inability to recognize something that is familiar, in spite of the intactness of the sensory system (sight, touch, smell, taste, sound) that underlies the observation. An example is the inability to recognize by touch a familiar object, such as a coin or keys placed in the palm, when the eyes are closed.

The agnosia relevant to this discussion is *anosagnosia* or *somatophrenia*. These terms refer to a condition in which an individual who has sustained a brain injury resulting in paralysis is unaware of the weakness or loss of function in the paralyzed body part. This symptom is associated with injuries of the right parietal cortex that result in left arm or leg paresis. Individuals with this symptom will state that their left side moves normally even when they are shown that the limb is paralyzed. Some individuals with anosagnosia believe that the paralyzed limb is not of their body but of someone else's. Others claim that they are able to move a paralyzed limb and are adamant that they are doing so, even when all observers agree that they are not. Still others "neglect" the part of the body that is affected—for example, do not shave part of the face or do not eat food on the part of a plate that is in the "neglected" visual field.

Thus, the agnosia syndromes suggest that an intact representation of the body in the sensory receptive area of the parietal lobe, particularly the right or nondominant parietal cortex, is one requirement for intact personal identity.

Peripheral Sensory Input

In a short autobiographical book, *A Leg to Stand On,* Oliver Sacks (1984) reports how an injury to a nerve in his leg led to the conviction and experience that his leg was alien, that it was not his own. What is intriguing is that Sacks had the cognitive awareness that his leg had been injured and that it was his leg, but nevertheless could not shake the conviction that it was not, because it did not "feel like" his leg. This account suggests that sensory input from peripheral sources, in this example from a leg, contributes to a construction of self, or at least to the experience of being a self.

The phantom limb syndrome might be considered the converse of the syndrome described by Sacks, but it also has its origin in injury (removal) of peripheral sensory input. In this condition, a digit or limb lost through amputation is experienced as still present. Individuals with the syndrome almost always have cognitive awareness that the limb is absent (just as Sacks had the cognitive knowledge that it was his leg and that it had been injured), but they nevertheless experience sensations from the limb that are convincing and imply that it is still present (a hallucination, because it is a sensation without an external stimulus). A similar experience of visual hallucinations associated with significant visual impairment (worse that 20/60) or loss of vision is referred to as the Charles Bonnet syndrome (Rabins 1994).

The absolute conviction with which these hallucinated experiences of body are reported, along with the importance that individuals ascribe to the mismatch between their cognitive knowledge and emotional valence, suggests that peripheral sensory input to the brain also contributes to the experience of personal identity.

Alteration of Personality

Traits are specific descriptive characteristics that can be observed in all human beings—that is, descriptors that identify universal characteristics of people. A large body of research supports the conclusion that traits are indeed universal (at least, they can be identified in individuals from all cultures studied) and that individuals differ from one another in the *amount* of any given trait. *Personality* is the constellation of these traits that is unique for each individual. Research demonstrating that personality is relatively stable throughout the life course in most individuals and that it can predict patterns of future behavior in groups of individuals supports the conclusion that *personality* identifies an abiding, important aspect of individuals.

Trait theory dates to the beginning of the twentieth century, although its roots are found in phrenologist Franz Joseph Gall's writings from the early nineteenth century (Hogan 1976). Two well-studied examples are the 2-factor model of Hans Eysenck that identifies primary dimensions of neuroticism (now called *stability* versus *instability*) and *introversion* versus *extroversion*, and the 16-factor model of Raymond Cattell. The most widely used model at present is a 5-factor model developed by Paul Costa and Robert McCrae that posits five primary personality dimensions: neuroticism (or emotional stability), extroversion, openness to experience, agreeableness, and conscientiousness (NEOAC), each of which is made up of other traits (McCrae et al. 2000).

Research demonstrates that personality is both heritable and environmentally shaped, with each explaining 30 to 50 percent of the variance. Clinical formulations of personality abnormality usually do not allow the diagnosis of a personality disorder to be made until patients reach the age of 15 or 16.

Several elements of the construct of personality are relevant to personhood. First, the notion that a set of unique and relatively unchanging descriptors can be applied to individual adults is an external validator of the idea of the uniqueness of individuals. Second, the universality of traits and the stability of personality, even in the face of significant environmental challenges, suggest that personality is embedded in relatively permanent brain structures. The universality of traits also suggests that evolution has played an important role in their range and establishment and that they emerge from brain systems that are genetically limited in their expression. Third, because there is also evidence that environmental influences are important in the final makeup of personality, it is likely that postnatal shaping of brain structure contributes not only to personality structure but also to personal identity.

Further cementing the relationship of personal identity to brain structure is the fact that brain injury can permanently alter the stable constellation of traits referred to as personality. The famous case of Phineas Gage—a railroad worker who was tamping an explosive charge that propelled a metal rod through his forehead, skull, and brain, and who underwent a dramatic personality change ("The balance between his intellectual faculties and animal propensities seems to have been destroyed")—is frequently cited as an early example (Harlow 1868). Since Gage, a large body of literature has accumulated linking brain injury to a dramatic and permanent change in what is considered the "essence" of an individual.

The location of Gage's injury, the inferior medial frontal lobe, is similar to the location of injury in many others who have had a personality change following brain injury. This suggests that those elements of personal identity referred to as *personality* depend on intact frontal lobes. There is no universal agreement on a more specific localization, but some evidence suggests that injury to the inferior medial frontal lobe results in a loss of inhibition—an increased expression of behaviors considered nonsocial—while injury to the more superior medial frontal lobe results in a more apathetic state.

Injury to this area of the brain is also characteristic of the frontotemporal dementias (FTDs), a group of progressive neurodegenerative diseases beginning in mid or later adulthood in which personality change or language disorder develops as an initial symptom of a progressive global dementia (Blass et al. 2004; Miller et al. 2001). Many individuals with FTD are brought to clinical attention because family and friends have observed them acting in ways that are clearly different from their usual behavior. While some individuals with FTD become primarily disinhibited and others primarily apathetic, all are recognized by family, friends, and acquaintances as having undergone an integral change in who they are. That is, their personal identity, as observed by others, has been altered. However, the individuals themselves, whether they have FTD or a frontal lobe brain injury, are usually unaware of the change and reject the claim that such a change has taken place.

The concept of *personality* identifies an aspect of personal identity that is observed in others and can be permanently altered by injuries or degenerative diseases of the frontal lobes of the brain. This suggests that intact frontal lobes are a necessary contributor to externally observed personal identity.

Memory

The necessity of memory for personal identity is demonstrated by the clinical condition referred to as Korsakoff syndrome, or amnestic syndrome. A widely studied example of this condition is the case of H.M., an individual who underwent two brain surgeries in the early 1950s to treat intractable epilepsy. During the surgeries, a pair of brain areas located deep in the right and left temporal lobes, called the (right and left) hippocampus, were removed because they were the "foci" or source of the seizures. Unfortunately, as became clear after the surgery, the presence of two hippocampi is necessary for the formation of new memories. After the surgery, H.M. could form no new verbal (word-based) memories—that is, he could not remember events that occurred in his life or any facts that he was presented with *after* the surgery, but he was still able to remember events that occurred before the surgery (Milner, Corkin, and Teuber 1968).

In essence, H.M.'s *auto*biography stopped when his hippocampi were removed. He lost the ability to recall anything that occurred in his life or in the world after the surgery, although he could fully comprehend, appreciate, and participate in events *as* they were occurring. However, to outside observers he was, until his death in 2008, the same, unique person after the surgery as before it; that is, his personality was unchanged. He remained able to speak normally, to interact with others as he had previously (i.e., his social manner was unchanged), to laugh at jokes, and so forth.

This lack of ability to form new memories robbed H.M. of the story of his life after the surgery (and of the knowledge that the surgery itself had occurred), and the knowledge of such events was integral to who he was and to his personal identity. In this instance, there was a mismatch between H.M.'s knowledge of his personal history (his "internal" personal identity) and the externally observed ("objective") personal identity known to others. We know, for example, that he had the surgery, that it changed his life trajectory, and that it resulted in his becoming dependent on others in a way that he had not been before. These facts seem crucial, at least to us, as external observers, in describing who he was—his personal identity. In this way, the inability to form new verbal memories impaired H.M.'s personal identity. What is intriguing is that his perception of his personal identity did not equate with that of external observers and that the genesis of this differing perception is known. The issue is not whether one view has primacy over the other, but rather that being able to form a personal history (memory of experienced events) is an element of personal identity *both* as it is experienced by the person and as it is observed by others. A mismatch between the observations of others and of the individual does not necessarily mean that the external observers are correct. However, in the case of H.M., the validity of the external observers can be upheld.

Furthermore, H.M.'s inability to form memories robbed him not only of his past but also of his future. That is, H.M., after surgery, was unable to plan for tomorrow or next week or next year because he could not put his current situation in the context of his personal identity. This future, or the ability to plan for the future, is referred to as *prospective memory* and is another aspect of memory that should be considered an element of personal identity.

The cases of H.M. and Gage, as well as the symptomatology of FTD, suggest that personal identity should include an element of external validation—the observations of others. Individuals can form incorrect memories that external observers believe are not part of that person's personal identity or that the person incorrectly believes are part of his or her personal identity, but it is the concurrence of the internal and external that makes us most sure that an accurate picture of an individual's personal identity has been formulated.

The Narrative Center

Another surgical operation performed to treat intractable epilepsy, the "split-brain" operation, can also result in the impairment of an aspect of personal identity. In this surgery, the corpus callosum, a band of white matter that connects the two cerebral hemispheres, is surgically split to prevent the spread of the electrical seizure activity from one hemisphere to the other. In a series of studies performed with patients who had undergone this surgery, Roger Sperry (who was awarded the Nobel Prize for this work) and his student Michael Gazzaniga were able to demonstrate that such individuals, and presumably all humans, have a center in the left hemisphere that strives to "make sense" of or develop narratives about events that are occurring. As reviewed by Gazzaniga in this volume (chapter 8), experiments can be devised in which information is presented to a single brain hemisphere. If some information is presented to the hemisphere that does not have primary control of speech (the nonspeech or "nonverbal" hemisphere) and different information is simultaneously presented to the other hemisphere (in which speech is processed), individuals will seem to "make up" stories that link the disparate information, even though they are unaware of the information presented to the nonverbal hemisphere. These elegant experiments suggest that human beings have a module in the left hemisphere that seeks to make sense of events by constructing a narrative that links the known facts together into a coherent whole. This center likewise participates in the construction, maintenance, and modification of a personal identity by linking together life experiences into a unique, comprehensive, and comprehensible narrative that we refer to as a personal identity.

This module for constructing stories remains intact in individuals who have the memory impairment of the amnestic syndrome discussed above. One intriguing aspect of the amnestic syndrome is that individuals will often tell stories that appear to be made up or preposterous explanations for events that seem impossible to other individuals. One hypothesized explanation for these "confabulated" stories is that the impairment of memory leads the person to put together disparate memories into a single narrative. They might do this because the left-hemisphere narrative center is intact and therefore pushing them to do so. While other individuals recognize the impossibility of the story, the person telling it is unable to do so because what is driving the behavior is the "push" by the narrative center to link memories, and the lack of memory impairs the person's ability to assess the accuracy of the linkages. While this is a hypothesized explanation, it does illustrate the plausible role for a narrative center in driving individuals to link information (memories) into a unique, meaningful construct that we label *personal identity*. Taken together, the amnestic and split-brain states suggest that the formation of personal identity requires both the ability to store verbal memories and the ability to link them into a coherent, comprehensive, sequenced, and unique narrative.

Summary

The neuropsychiatric approach demonstrates that multiple psychological constructs contribute to personal identity, that they are located in different areas of the brain (i.e., personal identity is a distributed capacity), and that personal identity can be altered by injury to specific brain areas. Each of these is a necessary aspect of personal identity, but none is sufficient to explain or delimit personal identity. A major criticism of the neuropsychiatric approach, however, is that it relies on the study of injured brains. Because the brain consists of multiple systems with many interconnections, it is likely that compensatory changes have occurred and that these explain or contribute to the phenomenon being studied. Thus, the neuropsychiatric approach cannot stand alone, and other methods must be used to confirm or disconfirm its hypotheses and conclusions.

Experimental Approaches

The study of performance in normal individuals offers another method for understanding brain function. Experiments to study the self and personal identity have been devised for 125 years, but the recent development of functional imaging methods, PET (positron emission tomography) scans and fMRI (functional magnetic resonance imaging), have revolutionized the experimental approach by allowing the identification of brain areas that are active when aspects of personal identity are being experienced or manipulated by an experimenter. These methods rest on the assumption that the activation of particular brain areas is accompanied by changes in metabolic activity of those areas and that these metabolic changes result in changes in blood flow that can be identified and measured by the scanning methods.

Recognition of Self

One set of experiments on the neural basis of personal identity examines whether the brain areas activated when individuals perceive their own face differ from those activated when observing the faces of others. Tilo Kircher, for example, demonstrated that self-recognition uniquely activates the left prefrontal cortex (Kircher et al. 2002). Similarly, in experiments with patients whose corpus callosum had been surgically cut, the left hemisphere was more activated in recognizing self, and the right hemisphere was more activated when recognizing familiar faces of others. (However, this finding has not been universally replicated; see Fossati et al. 2003; Keenan, Gallup, and Falk 2003.) If these study findings are confirmed, they imply that the left hemisphere plays a unique role in self-perception.

Another functional imaging study (D'Argembeau et al. 2005) found that the ventromedial prefrontal cortex, part of the area in which Gage was injured, was more activated when individuals reflected on their own personality traits than on those of others and that the amount of self-referential activity correlated with blood flow in this area. If this area is involved in modulating social behavior, as Gage's injury and subsequent behavior implies, then self-monitoring and self-perception might be distinct constructs that also contribute to personal identity.

Memory

Self-referential memory refers to the remembrance or identification of words or facts that are unique to an individual. Experimental manipulation of this capacity demonstrates that individuals remember facts or ideas that relate to themselves better than those that relate to others. Furthermore, brain areas in the left lateral prefrontal cortex that subserve language are activated in both conditions, whereas the medial prefrontal cortex is activated only by stimuli relevant to self-identification (Kelley et al. 2002; McCrae et al. 2000). The latter area was the site of Gage's injury, is among the areas damaged in FTD, and is the area damaged in brain-injured individuals who are impaired in the ability to self-reflect (Stuss and Alexander 2000).

Sensorimotor Contributions

Evidence supporting a sensorimotor contribution to personal identity, or "personal ownership," was demonstrated using the rubber hand illusion (Botvinick 2004). In this illusion, both a rubber facsimile of a hand, which the subject can see and is positioned where the subject's hand would naturally be, and the subject's own hand are repeatedly stroked. Eventually, individuals experience the stroking of the rubber hand alone as the stroking of their own hand. When the illusion occurs, areas in the right and left superior frontal cortex, the premotor areas, are activated. This occurs only when tactile (touch) and visual stimuli are simultaneous—that is, only when subjects look at the rubber hand and it is stroked simultaneously. The experiment supports the notion that the sensory system and input from the periphery are involved in the experience of one's physical body. It also demonstrates that the experience reported by Oliver Sacks (1984), as reviewed above, is dependent on brain-located mechanisms as well as peripheral sensory mechanisms.

Evidence from Development and Evolution

The capacity for self-recognition seems to be restricted to the great apes (Schilhab 2004) and to humans older than 14 months of age (Keenan, Gallup, and Falk 2003). Charles Darwin experimented with apes and with his own children and considered the recognition of self in a mirror as evidence of "higher intellect." Descriptive research from a variety of sources also suggests that humans begin to develop the recognition that they are separate individuals around two years of age and that any disruptions of the bonding and separation processes that take place at that time can have prominent adverse clinical repercussions in childhood and adulthood. Speculations on the evolutionary significance are reviewed by Churchland (2002).

These data support the inherent and thus genetically programmed origin of personal identity, as well as its importance in the psychological and interpersonal development of humans. However, self-recognition is one component of personal identity, not the equivalent, so other methods must be used to identify the origins and manifestations of the broader concept.

A Convergence?

The evidence from neuropsychiatry, experimental cognitive neuroscience, descriptive developmental studies, and comparative evolution strongly supports two conclusions. First, the construct of personal identity derives from the brain and can be disrupted by injury to the brain. Second, personal identity is neither in a single location nor a single capacity but, rather, it "involves a plurality of functions, each having a range of shades, levels and degrees" (Churchland 2002; see also Newen and Vogeley 2003).

Elsewhere in this volume, philosophers Marya Schechtman, Carol Rovane, and John Perry present their work on personal identity. Three ideas shared by all of these philosophers have parallels in the neuropsychiatric and experimental approaches reviewed here. First, they identify *narrative* as a core element. The left-hemisphere narrator identified by Gazzaniga and Sperry, the necessity of verbal memory, and the evidence of unique left-hemisphere activation in self-recognition all support personal narrative as a central contributor to personal identity. Second, the concept of *rational agency* championed by Rovane and also reflected in Schechtman's account of an external validating element has some parallels in the externally validated construct of personality, the experimental evidence of a unique self-monitor, and the demonstration that intact structure and function of the medial frontal lobe are necessary for the persistence of a unique personality—all of which supports the notion that there is a persisting, externally measurable contributor to personal identity. Finally, *intentionality*, the experience that one's personal identity is generated from within, is mirrored in several concepts discussed by the philosophers: self-reflection (subserved by the frontal lobes); the narrator, which seems to be organized to "make sense" of verbal input from the perspective of the individual and what he or she is perceiving; and body ownership, the sensory experience that one's body is unique and that one is able to interact voluntarily with the environment through movement and sensation.

This convergence of data from multiple sources and of philosophical constructs derived from introspection and the study of classical writings supports the validity of the construct of personal identity and identifies plausible generating mechanisms. Further study is needed to validate the existence and contribution of these mechanisms and to determine whether others, as yet to be detected, play a role. Because no single approach is best suited to elucidating a construct as complex as personal identity, future collaborations such as those presented here should be encouraged.

CHAPTER THREE

Case Studies

David M. Blass, M.D.

Case 1: Alzheimer Disease

Peter Jones is a 75-year-old man with the diagnosis of Alzheimer disease (AD). There is no history of dementia in Mr. Jones's family. He has no history of substance abuse. Before the onset of his illness, Mr. Jones had no history of psychiatric symptoms or treatment. He was a healthy child and a good student. He earned an M.B.A. from Harvard and went on to work for a Fortune 500 company, rising steadily through the ranks. He was never fired from a position and was considered an even-tempered, reliable worker. Mr. Jones retired at age 65 and traveled for a few years with his wife and friends, remaining active in social and community organizations. He has been married for 48 years and has three children and six grandchildren. He lives at home with his wife. His children describe him as a devoted husband, father, and grandfather who, until recently, relished time with the little ones, remembering their birthdays and always sending cards and gifts on time. He always managed the family finances conservatively.

Mr. Jones's medical history is significant for:

1. Elevated cholesterol
2. Type II diabetes mellitus
3. Hypertension
4. Osteoarthritis
5. Cataracts
6. Remote appendectomy

His current medications include baby aspirin, atorvastatin (for high cholesterol), metformin (for diabetes), metoprolol (for hypertension), multivitamin, and Tylenol as needed for pain.

For the first time, at age 71, Mr. Jones began having difficulty with his memory. His wife noticed that he would occasionally repeat a question just 5 to 10 minutes after asking it and would have no memory of asking it the first time. Although his driving remained safe, on a few occasions he made a wrong turn when returning home. These problems were initially attributed to "old age," as Mr. Jones did not appear much different from their friends of the same age. Mr. Jones continued to be active socially, working efficiently around the house and garden and exercising regularly.

Over the ensuing year, however, further changes were noted. Mr. Jones began to forget to send cards to the family for birthdays, and although Mrs. Jones helped him make up a calendar with the relevant information, he would forget to check the calendar regularly. Mrs. Jones noticed that he started to make subtle language errors, substituting *pencil* for *pen* and *table* for *desk*. He would begin home-improvement projects such as replacing light fixtures and then leave them uncompleted, a significant change from his baseline behavior. Despite this deterioration, Mr. Jones remained cheerful and optimistic, enjoying his time with family and friends and participating in community events. His friends, while noticing his deficits, continued to seek out his company, and he never lacked for activity.

Eventually, Mrs. Jones took Mr. Jones to see a neurologist, the daughter of a close friend. They spent the first 10 minutes "talking about the old days," and Mr. Jones remembered details from 40 years ago flawlessly. Mr. Jones knew he was there for an evaluation of his memory, though he did not feel that it was very impaired. When told that his family did believe his memory was impaired, he pleasantly responded, "Well, they usually know better than I do." Results of a general physical and neurological examination were completely normal. On cognitive testing, Mr. Jones was quite impaired. He did not know the date, he thought it was Wednesday instead of Friday, and he did not know which floor of the building they were on. He made errors in calculation. Object naming was fairly well preserved, but he made a few errors that were abnormal for his education level. Tests of new learning showed that he was very impaired. A brain MRI (magnetic resonance imaging scan) showed mild atrophy (shrinkage) throughout the brain without evidence of strokes. A diagnosis of probable AD was made about two years after the onset of the initial symptoms.

Although the Jones family was saddened by this diagnosis, daily life continued much as before, but the gradual decline in memory and functional ability continued. Memory-enhancing medications prescribed by the neurologist were only minimally helpful and were ultimately discontinued. At one point, Mr. Jones developed the belief that someone was hiding his wallet and keys from him, but this symptom improved when the family set up a system of putting these objects in the same place every evening. Mr. Jones had a bout of mild depression that was successfully treated with an antidepressant medication, which was eventually discontinued. Overall, Mr. Jones enjoyed his daily routine, and although he socialized somewhat less, he was generally content.

Two years after his official diagnosis, Mr. Jones was admitted to the hospital for treatment of hyperglycemia (elevated blood sugar). It became clear that he had not been taking his medications regularly. When this was discussed with him, he said that he thought he had been taking them properly but did not become angry or defensive about it. Later, he did not remember why he was in the hospital or what the problem was that led to the hospitalization. Mrs. Jones was educated about medication supervision and received assistance in devising a system to address this.

While he was in the hospital, it was noted that Mr. Jones had a large inguinal hernia that needed urgent repair to prevent intestinal damage. Mr. Jones agreed to the surgery, saying, "The doctors know best." He was able to state that if the surgery was not done, "something bad would happen," and he did not resist the preparations for the surgery. However, it was noted that 10 minutes after discussion of the surgery, he did not remember its purpose (other than "to help me") or its risks and benefits. A competency evaluation found him incompetent to consent to the surgery, and his wife took over this role. Mr. Jones interacted cheerfully and pleasantly with all of the hospital staff during his stay, complimenting and thanking them all, and had no inappropriate behavior.

Over the next three years Mr. Jones's condition deteriorated. He now required assistance in the bathroom and for showering and dressing. Although he no longer was embarrassed by these needs, he remained grateful to those who were assisting him in these matters. He remained appropriate in social situations, although he could not really follow the flow of conversation. He attended all family gatherings and even read a short speech at the wedding of their oldest grandchild. Because Mr. Jones could no longer travel, Mrs. Jones occasionally took trips with an Elderhostel group. Nonetheless, most of her time was spent with Mr. Jones, and they enjoyed being in each other's presence.

As the disease progressed, Mr. Jones could no longer recognize his grandchildren, although he did not mind their presence. He required 24-hour care because his gait had become somewhat unsteady and he had fallen a few times. He remembered Mrs. Jones's name only intermittently, but he still recognized her and was clearly comforted by her presence. Although her children on several occasions had gently encouraged her to consider moving Mr. Jones to a nursing home, out of concern for the toll that his care was having on her, Mrs. Jones refused to do so, feeling that he was clearly still benefiting from remaining at home.

• *Clinical Background*

Alzheimer disease is a brain disorder that affects as many as 4.5 million Americans. The disease usually begins after age 60, and risk increases exponentially with age. About 5 percent of men and women ages 65 to 74 have AD; as many as half of those age 90 and older have the disease. AD involves the parts of the brain that control thought, memory, and language. The disease is progressive and, in its most advanced stages, debilitating.

Alzheimer's begins slowly, with symptoms such as mild forgetfulness. In this stage, people may have trouble remembering recent events, recent activities, or the names of familiar people or things. As the disease progresses, people develop impairments in language (communication), perception (such as the ability to recognize familiar people and places), and the ability to perform everyday, learned activities (such as brushing their teeth or dressing). People with end-stage AD almost always need total care. The average person with AD lives 8 to 10 years after diagnosis, although some live as long as 20 years. The most common cause of death among people with AD is infection.

At present, treatment is symptomatic. Both medication and environmental therapies can improve quality of life, but whether disease progression can be modified is unknown.

The cause or causes of AD are not known. The most important risk factor is age, but scientists believe that genetics may play a role in some AD cases.

Case 2: Frontotemporal Dementia

Daniel Smith is a 55-year-old man who was diagnosed with frontotemporal dementia (FTD). There is no psychiatric or dementia history in Mr. Smith's family. Before the onset of his illness, Mr. Smith had no history of psychiatric symptoms or treatment. He was a healthy child and a good student. He earned an M.B.A. from Harvard and went on to work for a Fortune 500 company, rising steadily through the ranks. Before the onset of his illness, he had never been fired from a position and was considered an even-tempered, reliable worker. He has been married for 28 years and has three children, all in their early or mid twenties. He lives at home with his wife. There is no history of substance abuse. Family members describe Mr. Smith as, until recently, a devoted husband and father, sensitive to their needs and supportive in all ways. He was a responsible, steady person, active in communal causes and in the family's religious community. He always managed the family finances conservatively.

Mr. Smith's medical history is significant for:

1. Elevated cholesterol, successfully treated with dietary control
2. Borderline type II diabetes mellitus, also successfully treated for many years with dietary control
3. A remote appendectomy

He is taking no medications.

For the first time, at age 51, Mr. Smith began having difficulties at work. He frequently reported to his wife that his supervisor was unfairly critical of his work and that "people from Yale just don't like us Harvard grads." He would repeat this phrase almost every night, even when talking about their friends, family, or neighbors. Although he claimed that his work was still solid, his supervisor reported that he had become inefficient, condescending to others, and even rude on a few occasions, once even referring to a client's idea as "silly" (to his face). He would often be inappropriate in the work environment, making silly jokes and laughing loudly. After being given an ultimatum,

Mr. Smith agreed to be evaluated by a psychiatrist through the company's employee assistance program.

The psychiatric evaluation was inconclusive. There was no evidence of depression or mania, as the patient's mood, energy, sleep, appetite, and libido were all at baseline. There was no evidence of psychotic symptoms. Mr. Smith's memory, orientation, calculation, and visual-spatial perception were found to be normal, although he did get distracted easily. The psychiatrist speculated on the possibility of covert drug use, but a comprehensive toxicology screen was negative, as was a screening battery of blood tests. Ultimately, Mr. Smith was referred for interpersonal skills training, which he refused, and he was fired from the job.

Over the next year, Mr. Smith bounced from job to job, with the same types of outcome as above. He started eating significant amounts of sweets while at home, and this was attributed to his frustration at being unemployed, as was his increasing irritability. Over the ensuing months he began to take less interest in his appearance, often refusing to shave even for social engagements. He insisted on wearing the same clothes day after day, and Mrs. Smith found herself surreptitiously washing his clothes at night to prevent them from smelling. His conversation at home became narrowly focused on food, his inability to obtain a job because of "those Yale guys," and TV sitcoms. He took less and less interest in the activities of his wife and children. Although they initially attributed all of this to the stress and embarrassment of being unemployed, family members began to suspect that something more was wrong, but out of respect for him, they did not initially confront him. When they did raise their concerns, he dismissed them sarcastically.

Mr. Smith often walked the streets during the day. Although he would never get lost, he began to get into frequent trouble. On several occasions he made racial slurs when walking past people on the street. He was arrested for kicking a dog sitting on the curb, and it required the intervention of the family attorney to get him released. He no longer was able to go to social events because of his rude and inappropriate comments. At his daughter's wedding, he had to be given a peripheral role because of concern about his spoiling the occasion.

Mr. Smith began to gain weight. He had several hospitalizations for treatment of hyperglycemia (elevated blood sugar), but would leave the hospital as soon as he was stabilized and then refuse to take medications. A competency evaluation found that although he seemed to act impulsively and in poor judgment, he could remember and apparently understand the information presented to him, including the details of the treatment options and the consequences of not receiving treatment. When asked why he was acting this way, he stated, "You only live once, I might as well enjoy myself while I'm still here."

Over the next few months, Mrs. Smith noticed that Mr. Smith began having difficulty thinking of the words he wanted to say. Although his comprehension seemed to be intact, his verbal output became more halting and his sentences became shorter. At this point she insisted that he be evaluated by a neurologist. The neurologist noted that Mr. Smith appeared disheveled, silly, and sarcastic at times. Mrs. Smith described the other changes to the neurologist and told her, "It is as if a different person has crawled into Daniel's skin." His speech was halting and he had difficulty naming simple objects. He had absolutely no insight as to why he was at the neurologist's office. Results of a comprehensive physical and neurological examination were normal. On cognitive testing, once again his performance was within the normal range for orientation, calculation, and visual-spatial perception. He had difficulty with new learning, but it was not severely impaired. Performance on tests of judgment, abstraction, and executive function was relatively worse than in the domains listed above. A brain MRI scan revealed atrophy (shrinkage) of the right frontal and left temporal lobes. The diagnosis of FTD was made.

As time progressed, Mrs. Smith developed a routine of her own, independent from Mr. Smith. She attended religious services and social events without him, arranging for a care provider to be with Mr. Smith while she was away. Not only could Mr. Smith no longer participate in decision-making for the family, but Mrs. Smith no longer shared with him the issues she was contemplating or the results of her decisions.

Over the next three years, Mr. Smith continued to decline gradually. He remained at home most of the day, watching television. His conversation was centered almost completely on food. He reacted without any feeling to those around him, including to the news that his wife had been diagnosed with a malignancy. When the news of the tragedy of September 11 was displayed on the television, he reacted with annoyance that his programs had been interrupted. When news of the birth of his first grandchild arrived, he was transiently joyous, then immediately returned to his TV show.

Although Mr. and Mrs. Smith had always told each other that one would try to care for the other at home in the event of a protracted illness, even dementia, Mrs. Smith eventually found that having Mr. Smith at home was unbearable. Not only did he require 24-hour supervision for safety purposes, but the time they did spend together was no longer meaningful. He did not seem to enjoy her company and seemed indifferent to her (or anyone else's) presence. She began to consider placement in a nursing home.

• *Clinical Background*

Frontotemporal dementia is increasingly recognized as a fairly common type of dementia, and its clinical course differs significantly from that of Alzheimer disease. FTD is a degenerative condition primarily affecting the frontal and anterior temporal lobes. These areas primarily control judgment, personality, regulation of behavior, movement, speech, social graces, language, and some aspects of memory. The age of onset of FTD is typically in the fifties and sixties, with equal incidence in men and women. In about one-third of patients, a family history of dementia exists, suggesting a genetic component in the etiology of many cases.

This type of dementia is marked by dramatic changes in personality, behavior, and judgment. Changes in personal and social conduct occur in early stages of the disease and include disinhibition, socially inappropriate behavior, apathy, social withdrawal, hyperorality (greatly increased food intake and mouthing of nonfood objects), and ritualistic compulsive behaviors. Profound lack of regard for personal hygiene is common. These symptoms may lead to misdiagnosis as a mood disorder, personality disorder, or the product of familial conflict, while in the elderly they may be mistaken for withdrawal or eccentricity. Family members often describe these patients as having become "a completely different person." FTD patients often present two seemingly opposite behavioral profiles in the early and middle stages of the disease. Some individuals are overactive, restless, distractible, and disinhibited, while others are apathetic, inert, aspontaneous, and emotionally blunted. Early in the disease, cognitive functions such as memory, calculation, and orientation may be preserved, but eventually, the clinical picture progresses to one of significant cognitive impairment.

Frontotemporal dementia can be diagnosed with relative accuracy by completing a careful neuropsychiatric history, supported by neuroimaging and neuropsychological testing. The length and progression of FTD vary. Some patients decline rapidly over two to three years, while others show only minimal changes over a decade. No medications are known to treat or prevent FTD, but serotonin-boosting or dopamine-boosting medications may help alleviate some behaviors.

Case 3: Deep Brain Stimulation

The year is 2012, and deep brain stimulation (DBS) is commonly employed in the treatment of patients with Parkinson disease (PD) in the United States. DBS has been found to improve the motor symptoms of PD (tremor, rigidity, and gait disturbance) in many patients, and interest developed in using DBS to treat some of the psychiatric complications of PD.

With further refinement in technique, it was discovered that placing a fairly large number of electrodes deep into the brain can sometimes result in improvement in severe apathy. The only drawback is that in contrast to traditional DBS, in which the treatment is reversible, this modified DBS technique is not.

Charles Garrison is a 61-year-old man with moderately advanced PD who was referred for DBS treatment. Before developing PD, Mr. Garrison was a successful engineer for a military research firm. He rose through the ranks to become the director of aeronautics research, earning a reputation for diligence and conscientiousness. He was quiet, matter-of-fact, and somewhat shy, but always energetic and enthusiastic about new ideas at work. He worked six days a week and had few friends. He was married with three children and spent all of his free time with his family, who shared his enthusiasm for ideas and experimentation.

After developing PD, Mr. Garrison initially continued to work. His company was accommodating of his need for a flexible schedule, frequent absences, and modified workspace. As the disease progressed, Mr. Garrison developed a fairly severe apathy syndrome that impaired his job performance. He no longer initiated new projects, began to miss deadlines, and often did not show up for work at all. At home, his family described him as "a changed man," no longer showing interest in his children, skipping their school performances, and responding to their stories with minimal emotion. He no longer "tinkered" in his shop and stopped reading the newspaper. He denied feeling sad and did not express feelings of hopelessness or guilt. His sleep and appetite were normal. His neurologist diagnosed an apathy syndrome related to PD, and a psychiatric consultation gave a concurring result, finding no evidence of depression or cognitive impairment. Mr. Garrison was offered the newer technique of DBS as a potential treatment for both the motor symptoms of PD and the severe apathy. After deliberation about the risks and benefits and after obtaining a second opinion, the family decided to proceed with the DBS. Although they were concerned about the irreversibility of the modified DBS technique, they felt that the potential benefits of reversing the severe apathy outweighed the risks.

Following implantation of the DBS, Mr. Garrison had significant improvement in his motor symptoms. More dramatic, however, was the change in his personality and demeanor. Previously shy and introverted, Mr. Garrison now became extremely outgoing and gregarious. He would seek out crowds of people to speak to and frequently became the center of attention. He spent hours at work telling stories instead of working on his projects. At home, during meals, he redirected the topic of conversation to something he could dominate. Although he demonstrated an apparent renewed interest in the happenings of his family, the interest went only as far as being able to attract the attention of the group onto himself.

Mr. Garrison went on to develop not only a new demeanor but also a new outlook on the world. Previously a loyal Republican, he switched his affiliation to the Democratic Party. He became an ardent environmentalist, traveling to numerous conferences and insisting (over his wife's objection) on giving all of their charity donations to environmental causes. Over time his interests shifted to a variety of other social, political, and charitable causes, and he threw himself passionately into each one in turn. Mr. Garrison decided (without consulting with his wife or children) to quit his job so he could devote more of his time and energy to these various causes.

• *Clinical Background*

Apathy is a clinical syndrome commonly found in neuropsychiatric disorders. Loosely defined, *apathy* refers to a significant diminution of motivation, feeling, emotion, interest, or concern that is not purely due to a decreased level of consciousness, cognitive impairment, or emotional distress (Starkstein et al. 2001). Although patients with clinical depression also appear apathetic, the apathy syndrome described above is defined even in the absence of frank clinical depression.

The core feature of the apathy syndrome is impairment in motivation and interest. This can encompass all or most aspects of life, including work, family, social and religious life, and so forth. Interest in food often remains intact.

Some patients retain an interest in activities or people, yet have no apparent motivation to act on this interest. Some patients are cognizant of this discrepancy and baffled by it. They say that they do not know why they are unable to carry out a task; they just are unable to do it. Apathy can be severe, with some patients essentially remaining in bed all day, without getting out even to use the bathroom. Others lose interest in family members and may react without emotion on learning that a close relative has just been diagnosed with a fatal disease.

The apathy syndrome is prevalent in many neuropsychiatric diseases, including AD, PD, FTD, stroke, traumatic brain injury, and normal-pressure hydrocephalus. The apathy syndrome may improve modestly after treatment with medications that increase dopamine in the brain and may also improve with structured behavior-modification plans. This syndrome can be among the most troubling areas for family members caring for patients with these disorders.

Case 4: Steroid Psychosis

John Fast is a 26-year-old professional athlete. He has been active in athletics since junior high school, consistently well regarded by his teammates and coaches. Throughout his career, despite his overwhelming success, Mr. Fast has always been a "team player." During an interview several years ago, his coach referred to him as the "cohesive factor" for the team, and several teammates have gone on record describing him as "modest" and "a role model for young athletes." Since high school, when his talent first began attracting attention, Mr. Fast has always had an excellent relationship with the press. In one article, a reporter described him as "playful, but polite and courteous. A real joy to interact with." Mr. Fast has been married for five years and has a two-year-old daughter, whom he adores. Until fairly recently, his relationship with his wife, Jenny, was excellent. In fact, the happiness they radiated often inspired playful exclamations of jealousy from friends and family members.

Two years ago, Mr. Fast surreptitiously began taking steroids to improve his strength. His career had thus far been successful, but he wanted to "give it a boost." He had no prior psychiatric history. After three months of steroid use, he developed the belief that he was the strongest person in the world. He began spending lavishly and then developed the conviction that he was being persecuted by the team general manager because of jealousy and other players' inability to accept his superiority. Mr. Fast's behavior became increasingly erratic and his performance declined dramatically. At home, he began yelling at his daughter with little or no provocation, and his relationship with his wife began to decline. When she complained that he was being selfish and behaving poorly, he claimed that she was simply jealous of his success and intimidated by his superior strength. After a few weeks of this behavior, his coach insisted that Mr. Fast undergo a psychiatric evaluation, and a diagnosis of substance-induced (steroid) bipolar disorder, manic phase (also referred to as "steroid psychosis"), was made. After treatment and cessation of steroid use, Mr. Fast returned to baseline.

One year later, Mr. Fast began worrying that his performance was declining and he began taking steroids again. He initially performed well, but as the season progressed he became irritable, demanding, and aggressive with teammates. He insulted the team owner and general manager, and these verbal insults were quoted in the newspaper. At Mr. Fast's insistence he was traded, even though his wife opposed the plan. At home he was often irritable and argumentative. Jenny's sister even suggested that their once enviously healthy relationship was becoming verbally abusive. These changes were initially attributed to the stress of adapting to a new city and team, but the pattern persisted.

On his new team Mr. Fast continued to get into arguments with teammates and eventually the general manager, and he was described as destructive and egocentric. Even though the team made the playoffs, he was suspended because of behavior detrimental to the team. He was then referred to the team psychiatrist, who diagnosed him with steroid-induced mood and personality change.

Ending 1

Despite the diagnosis of steroid-induced mood and personality change and the repeated pleas from his coaching staff and teammates, Mr. Fast refused to stop taking the steroids. He believed that he was a better athlete and a better person because of the steroids, and he told his psychiatrist that he liked his "new self" better than the old one. His behavior continued to be erratic and irresponsible. His coach suspended him from play, and at the end of Mr. Fast's term his contract was not renewed. When he refused to stop taking the steroids and also refused to undergo marital counseling, his wife moved out of their home and took their daughter with her.

Ending 2

After his diagnosis, Mr. Fast consulted with his coach and received reassurance that, despite sacrificing some of his newfound strength, his career would be much better if he quit taking the steroids. His wife and several of his teammates also asked him to stop using the steroids. Two weeks after his diagnosis, Mr. Fast decided to stop taking the drugs, and his behavior shortly returned to baseline. When confronted with stories of his behavior while taking steroids, he said he was "embarrassed and ashamed" by the way he acted.

• *Clinical Background*

Steroids are known to cause a variety of neuropsychiatric symptoms. In the affective realm, they can induce both major depression and mania (bipolar disorder, manic phase). In the cognitive realm, they can induce delirium, an acute, acquired impairment of cognition accompanied by an altered level of consciousness. Steroids can also induce persecutory (paranoid) delusions without alteration in mood or cognition.

The term *steroid psychosis* refers to the fact that steroids can induce hallucinations and delusions. These symptoms can be associated with an affective disorder or a cognitive disorder, or they can occur in isolation. The mood disorder sometimes persists even when the steroid medication is discontinued; however, when these symptoms occur as part of a delirium or isolated persecutory delusions, they usually resolve when drug use is stopped.

In this case, the patient seems to have a steroid-induced mood or affective disorder. The symptoms referred to as *psychotic* are delusions congruent with the diagnosis of bipolar disorder, manic phase.

PART II / Philosophers Hold Forth

CHAPTER FOUR

Getting Our Stories Straight

Self-narrative and Personal Identity

Marya Schechtman, Ph.D.

In everyday life we have relatively little trouble reidentifying people. Most questions of personal identity that we encounter can be settled by establishing relevant facts about human bodies. I may be uncertain that the person I am picking out of the police lineup is really the same person I witnessed running from the crime scene, but spying a distinctive tattoo may help settle the question for me. If I fear that the person claiming to be my long-lost brother is really a stranger trying to defraud me, I may ask him to take a DNA test; and if I have trouble distinguishing between the identical twins who work for me, it will be helpful to learn that one of them has a scar over her left eye. We may not always be able to obtain the information we need to resolve these sorts of identity question decisively, but we know what kind of information would enable us to do so.

Sometimes, however, we encounter a different sort of question about personal identity. In these cases, we are confident that we are dealing with a single human being, but that human being shows psychological alterations so profound that we question whether what we are encountering is really a single *person*. This kind of identity question might arise, for instance, in dealing with someone with dissociative identity disorder (DID) who seems to exhibit several distinct personalities.[1] It also arises in the four case studies we are asked to consider. Each of these cases describes a human being who changes in such fundamental ways that it is natural to ask whether we are dealing with the same person throughout his story. These identity questions cannot be answered by learning more facts about human bodies, because the sameness of the human being is not at issue. With these questions of identity, unlike the ones noted above, it is not immediately obvious how to go about resolving them. My primary goal in this chapter is to get clearer on how these kinds of question *can* be addressed.

I begin by suggesting that the questions of personal identity that arise in cases such as DID and our case studies are concerned with what I call *forensic personal identity.* This is the notion of personal identity that underlies important practical issues concerning moral responsibility and entitlement. After defining the notions of forensic personal identity and *forensic personhood,* I briefly discuss the way in which the four case studies raise questions about these concepts. Next I sketch my account of forensic personal identity, which I call the *narrative self-constitution view,* and show how this view can be fruitfully applied to the specific questions raised by the case studies.

Forensic Personhood and Personal Identity

The first order of business is to see that cases such as DID and our four case histories raise questions of personal identity. It is in some ways tempting to conclude that the "questions" raised here are not genuine questions about personal identity but merely rhetorical devices used to express an anomalous degree of change in a single person. It seems clear that we sometimes use questions of personal identity in this way. I might say of someone, for instance, that she is "a different person" since taking a vacation or finishing a stressful long-term project. If someone is severe professionally and compassionate personally, we might say that she is "a different person" at the office and at home. In either case, these sentiments might be expressed in a question: "Are you really the same person who was biting everyone's head off while you were trying to finish that report?" or "Can this charming hostess really be the same person who reduces everyone in the office to tears on a regular basis?" In these cases the questions do not seem to signify any real puzzlement about identity, but instead are a registering of how multifaceted people can be.

One possibility, then, is to say that DID and Alzheimer disease, frontotemporal dementia, steroid psychosis, and the deep brain stimulation described in our case studies are continuous with these more common cases. On closer analysis, however, we can see that the questions raised in these more extreme cases are of a different sort and that their answers have implications. To see this, we must appreciate that facts about personal identity are of immense practical importance. Whether, for instance, you are the same person who made me a promise determines whether I have a right to expect you to fulfill it; and whether you are the same person to whom I made a promise determines whether you have a right to expect fulfillment from me. Whether you are the same person who did work for me determines whether you are the person who is entitled to the compensation for that work; and whether you are the same person who committed a crime is vital to determining whether it is fair to punish you for it. Many of our practices concerning responsibility, compensation, and entitlement depend, that is to say, on facts about personal identity. Of course, these facts are not always the final word on such questions. Even if I am the person who made you a promise, you may not be entitled to demand that I keep it if I made it only because you tricked me; and even if you are the person who acted in a way that was harmful, you may not be held responsible for the action if it was done unknowingly. Still, it seems as if settling questions of identity is a starting point for asking these more subtle questions about entitlement and responsibility, and so identity plays a particularly fundamental role in these practices.

Cases such as DID reveal the possibility that the limits of the person, for these purposes, may not coincide with the limits of a single human being. When we ask, for instance, whether the different personalities displayed by an individual who has DID really represent distinct persons inhabiting a single human body, we are asking, among other things, whether these personalities should be treated as separate persons for the practical purposes described above. We wonder, genuinely, whether the different personalities can be held responsible for actions undertaken when an alternate personality is controlling the body, or whether the alternate personality now in front of me is bound by the promise made by this human being but by a different personality. As we shall see, similar issues are raised by the four case studies. These identity questions are quite different from the more rhetorical identity questions raised above. We may judge that the responsibility of the woman who is nasty while completing her project should be mitigated by the fact that she was under pressure, but there is no question that, from the standpoint of assessing responsibility, the nasty actions are hers. In the case of DID, however, it is not just that we think someone's responsibility may be mitigated by the fact that, at the time of the action, an alternate personality controlled her body; rather, we think that, for purposes of determining responsibility, it may not be properly considered her action.

There seems, then, to be a notion of personal identity that serves as a minimum condition in the assessment of responsibility, obligation, and certain sorts of entitlement. Our practices of promising, contracting, and assessing praise or blame depend on this notion. In cases in which there are radical psychological discontinuities within a single human life, questions arise as to whether this human being should, for the purpose of these practices, be considered one or more than one person. A great deal that is important to us turns on this. John Locke, who in many ways set the terms of current philosophical debates about personal identity, says that *person* is a "Forensick term" (Locke 1975).[2] In the spirit of Locke, I will call the questions of personal identity raised in cases such as DID questions of *forensic personal identity.* Cases of this kind raise a question about the conditions of forensic personal identity, intimating that these conditions may not be determined by the conditions for the identity of a single human being.

Before discussing questions of forensic personal identity as they arise in our four case studies, I should point out that the idea of forensic personal identity naturally implies a notion of *forensic personhood.* To be a person in the forensic sense (a *forensic person*) is to have the capacities to engage in the practices in which judgments of forensic personal identity are relevant. Forensic persons, that is, are those who have the psychological capacities to act as moral agents and to enter into binding contracts and commitments. This means that not all human beings will be forensic persons, nor will all forensic persons necessarily be human beings. Those humans who lack these capacities will not count as forensic persons, while primates or aliens who had them would count as persons in this sense.

Many will be understandably nervous about a view in which some human beings are nonpersons or not fully persons. There is always the worry that these claims will be used to justify ill-treatment. I am not sure that I can lay these worries fully to rest, and I do believe that we must remain constantly vigilant with respect to them. It is essential to understand, however, that the implications of judging that someone is not a forensic person are relatively circumscribed, even if they are deeply important. The judgment that someone is not (or not fully) a forensic person relieves him of certain kinds of responsibilities and commitments. There are certain behavioral expectations we can have of forensic persons, and when they fail to live up to those expectations we often are justified in censuring them for being irrational or immoral. To say someone is not fully a forensic person is to say that this kind of censure is inappropriate. To take a somewhat trivial example, many may find it somewhat chilling to say that a human infant is not a person, but few will deny that it is inappropriate to form a negative moral judgment about that infant because he so inconsiderately interrupts our conversations and sleep with his crying.

The failure to be a forensic person also has an effect on a person's entitlements, but only on specific entitlements. Many different facts about us determine how we deserve to be treated and, of course, there is a great deal of disagreement about just what our entitlements are and how we come to have them. Most would agree, however, that a creature is entitled to certain considerations just by virtue of being sentient. We generally believe that the gratuitous infliction of suffering on creatures capable of suffering is wrong. Many also believe that there is a further set of entitlements that we have by virtue of being human. We do not allow people to keep humans as pets, or euthanize them, or sell them if they no longer wish to care for them, or kill them for food. This culture, at least, does allow us to treat many sentient animals in this way. Others argue that living things are entitled to certain considerations just by virtue of being alive, though what they are owed is generally thought to be somewhat less than what sentient creatures are owed.

While there is, as I have said, no uncontroversial understanding of what kinds of consideration different creatures are owed, the important point here is that none of what *is* owed on this score (whatever that turns out to be) is lessened or negated by the failure to be a forensic person. If someone is not a forensic person, this does not imply that we can be indifferent to her suffering, or terminate her existence for our convenience, or view her as an object. She is still a human being and entitled to all of the considerations due to human beings. The entitlements that go with forensic personhood are of a different sort. For the most part, they are entitlements to certain kinds of self-determination. In the case of some human beings—again, take infants as a relatively uncontroversial example—we believe that the capacities necessary for successful self-determination are not present, and so we do not allow infants to determine the course of their own existence. In our dealings with them we are unapologetically paternalistic, focusing on their well-being and not worrying about their autonomy. We tend to believe that—or at least think we must consider whether—this is also the appropriate attitude to take toward those who cannot develop capacities for self-governance much beyond those of infants, or who develop such capacities and later lose them.

Forensic personhood, as I have defined it, is a good for us. In general, we believe that those who, for one reason or another, never develop into forensic persons are missing something worth having and that it is a misfortune to be in this condition. This does not mean that forensic personhood is the only or the highest good. It does not mean that the lives of those who do not develop into forensic persons are less valuable than those of forensic persons, or that it would have been better for them not to have been born. It does not even mean that, all things considered, it would necessarily have been better for them not to have the conditions that prevent them from developing into forensic persons. Overall, the life of a nonforensic person might be better and contribute more to the world than the life of a forensic person. What is true, however, is that those who cannot develop into forensic persons are prevented from entering into certain sorts of interactions and practices that we value highly. The kind of value we place on forensic personhood can be seen in the way most forensic persons respond to learning that they have a disease that will ultimately rob them of the capacity for forensic personhood. Forensic persons who receive such diagnoses usually find this prospect horrifying and depressing. The loss of the ability to be a self-determining agent, someone who can be held responsible for what he says and does, is profound and not trivial.[3]

Of course, the capacities associated with responsibility and self-governance admit of degrees, and so there will be degrees of forensic personhood as well. We rarely, if ever, encounter an adult human being who is a forensic person to *no* degree. What we typically find instead are cases in which forensic personhood is compromised, in some particular way, to a greater or lesser degree. Because we value forensic personhood, we typically believe that whatever degree of forensic personhood someone possesses should be nurtured and respected. This is why we treat 3-year-olds differently from infants, 7-year-olds differently from 3-year-olds, and 14-year-olds differently still. It is considered an insult—and perhaps an injury—to act as if someone cannot appropriately be held responsible for his actions or bound by his promises when in fact he can. This means that in cases in which questions of forensic personal identity arise, we are interested in determining to what degree and in what ways someone *is* a forensic person as well as wanting to know whether he is the *same* forensic person over time. As we shall see, both sorts of question often arise at once.

Questions of Forensic Identity in the Four Case Studies

The four case studies we have been asked to discuss are all situations in which issues of forensic personhood and personal identity become extremely complicated. In each case there is some sense in which forensic personhood and/or identity is compromised, yet some degree of personhood and identity is retained. These cases are likely to invoke some pretty clear-cut judgments of forensic personhood and identity, and also some confusion and questions. In this section I characterize likely responses to these cases, focusing on issues of forensic personhood and personal identity. When we have clear reactions to these cases, these can be used as data to develop an account of these concepts; when we are confused about these issues, we find questions that a theory of forensic personal identity should be able to answer. A discussion of initial reactions to these cases will thus help focus discussion and aid in the development and evaluation of an account of forensic personal identity.

Let's begin with the case of Mr. Jones, who has Alzheimer disease (AD). This disease seems to attack forensic personhood itself more than it interferes with forensic personal identity. We are less tempted, that is, to say that Mr. Jones is turning into someone else than to say that he is gradually fading away, becoming less and less capable of interacting with the world and with others in the ways he used to. To the extent that we think Mr. Jones should not be held responsible for actions and promises in the later stages of his illness, it is not because we think he is not, in the relevant sense, the same person who took those actions or made those promises, but rather because he is losing the capacities that make him the sort of agent who can be held responsible in this way. Even if it was always Mr. Jones's job to send the cards and gifts to the grandchildren, for instance, and even if he promised before his illness that he would make sure birthday tokens arrived on time, it would not make sense to take him to task for not living up to this promise when his impairments do not allow him to even remember having made it. This is not because it is not *he* who made the promise, but rather because he is no longer able to keep it. As the disease progresses, it becomes increasingly difficult to treat Mr. Jones as a forensic person. The fact that he is found incompetent to consent to his surgery underscores this change.

Some interesting questions about forensic identity do arise in this case, however, when we consider the position of Mrs. Jones. She has made, undoubtedly, many promises and commitments to Mr. Jones, including those that are part of marriage. While we may feel clear that it is inappropriate to expect Mr. Jones to keep his commitments to Mrs. Jones, it is less clear to what extent—if any—Mrs. Jones is no longer bound to the commitments she made to Mr. Jones. Is he, in any relevant sense, no longer the person to whom she made these promises? We can see by looking at this case story that Mrs. Jones feels strongly bound to her fundamental commitments to care for Mr. Jones's well-being. She maintains her relationship with Mr. Jones to the extent possible and refuses to put him in a nursing home. This attitude seems clearly the right one to take. This is not to say that it would be obviously wrong to put Mr. Jones in a nursing home. In the right circumstances, that might be the most loving decision and the best way to fulfill Mrs. Jones's spousal commitments. It does seem obvious, however, that something would be wrong with Mrs. Jones deciding that this demented Mr. Jones before her is not, for forensic purposes, the same person to whom she made her vows and that she therefore can wash her hands of him and need have no interest in how he lives out the rest of his life.

It will be useful to get a bit clearer on the kind of obligation Mrs. Jones has to Mr. Jones and how it is "forensic." To a certain extent, of course, Mrs. Jones should be interested in what happens to Mr. Jones just because he is a sentient human being, but her obligation clearly goes beyond this. She owes him consideration that others who can equally well recognize his humanity do not. She has this obligation, to state the obvious, because Mr. Jones is her husband; they have a history, and they have made promises to each other. It is important to be clear here that by using terms such as *promise, vow,* or *obligation* I do not mean to imply that Mrs. Jones should care about Mr. Jones simply out of a cold sense of duty or because some law—civil or moral—requires it of her. Many of the promises we make to one another are made in love or friendship or collegiality, and they are often kept in this spirit. Much of what I am putting under the rubric *promise, commitment,* or *obligation* is not a formally spoken contract but rather refers to the ways in which we come to owe certain sorts of consideration to those with whom we have a history. I do not need to tell my best friend and confidante since early childhood that she may ask for help from me in times of crisis; that is the relationship we have. The point I am making here is that a host of reciprocal commitments usually evolve in interactions between forensic persons, and this case reveals the strong sense that these commitments do not simply dissolve when the person with whom one is in a relationship becomes unable to reciprocate. There may be some relationships that are appropriately dissolved in these circumstances, but close personal relationships such as the spousal relationship are not among them. It thus seems clear that Mrs. Jones should view Mr. Jones as the same person for at least these forensic purposes.

There are, however, more subtle issues about forensic identity that are not so easy to answer. As Mr. Jones sinks farther into dementia, a tension may develop between acting to promote the flourishing of Mr. Jones before the onset of AD and acting to promote the flourishing of the present Mr. Jones. The earlier Mr. Jones, for instance, might have been mortified at the thought of his grandchildren seeing him in his current condition, while the present Mr. Jones seems to get some pleasure from their visits even though he does not recognize them. If we assume that Mrs. Jones's commitment is to the person she married, then the question is whether that commitment is better honored by giving priority to the wishes of the earlier or the present Mr. Jones. If we conclude that the Mr. Jones with late-stage dementia is a different forensic person than the earlier one, then Mrs. Jones's primary forensic responsibilities are to the predementia Mr. Jones; but if he is the same forensic person, then these responsibilities are to the postdementia Mr. Jones. Unfortunately for Mrs. Jones, she is likely to feel that she is betraying her husband whichever path she chooses.

Consideration of the case of AD thus provides the following insights and questions about forensic personhood and personal identity. First, Mr. Jones clearly shows a continuously decreasing level of forensic personhood, and any theory of forensic personhood should account for that fact. At the same time, he never entirely loses his forensic personhood, and that should be explained as well. It seems clear from this case that he is the same forensic person to the extent that Mrs. Jones should still view herself as committed to him in the way one member of a couple is committed to the other. This should also be an implication of a viable account of forensic personal identity. Finally, we encounter some more difficult questions about whether to treat Mr. Jones as a single person for forensic purposes for cases in which the interests of the pre-

and postdementia Mr. Jones seem to be very much at odds. It would be desirable for an account of forensic personhood to provide some insight there.

Next we turn to the case of Mr. Smith, who has frontotemporal dementia (FTD). The discussion can be briefer here because much of the groundwork for this and the next two cases was laid in our discussion of AD. It will be illuminating, however, to focus on some important differences between this case and the case of Mr. Jones. Mr. Smith's impairments are primarily affective rather than cognitive. Compared with Mr. Jones, he retains a great many of his faculties and capacities, but his forensic personhood is severely compromised nonetheless. Mr. Smith, through his lack of concern for himself and others, becomes totally unsuited to making important life decisions or engaging in the kinds of relationships and interactions typical of adult human life. For the most part, it does not seem appropriate to hold Mr. Smith responsible for his behavior. His racial slurs, cruelty to animals, inappropriate social behavior, and indifference to his wife's malignancy are evidence of his illness or diminished capacity, not evidence of an evil nature. This case thus shows that the capacities necessary to forensic personhood are not just cognitive but affective as well. The incapacity to care also diminishes forensic personhood.

Despite the profound effect of Mr. Smith's dementia on his forensic personhood, however, he retains enough personhood that it is more natural to describe his case as a *change* in forensic identity than it was for Mr. Jones's case. Although Mr. Smith has few interests and concerns as his illness progresses, he is able to make his wishes known in a clearer and more forceful way than could Mr. Jones toward the end of his illness. Mr. Jones's ability to participate in ordinary human interactions diminishes drastically, but at each stage, insofar as he is able to interact, his behavior is appropriate to his circumstances and continuous with his earlier life. He remains affectionate, compliant, and considerate to the extent that he is able. The questions about his forensic identity arise when his abilities have declined so far that his desires start to look discontinuous with those of the previous Mr. Jones rather than like truncated versions of his earlier desires, as they did earlier in his illness. Mr. Smith, by contrast, expresses desires and attitudes radically discontinuous with his earlier ones. This is what brings Mrs. Smith to say that it is "as if a different person has crawled into Daniel's skin."

Mr. Smith's focus on television programs and food and his neglect of personal hygiene are undoubtedly symptoms of a pathology. This pathology is coupled, however, with a much higher level of forensic personhood than we saw late in Mr. Jones's illness. The presence of a relatively high level of forensic personhood despite the dementia makes it somewhat clearer in Mr. Smith's case, compared with Mr. Jones's, that the predementia Mr. Smith's view of his current life does not warrant interference with the postdementia Mr. Smith, who is indifferent to his own poor hygiene, for example. Mr. Smith is enough of a person to make decisions about his life despite his impairments, and this is echoed in the findings of the competency evaluation conducted at the hospital. To nurture the forensic personhood of Mr. Smith in the later stages of his dementia, then, he must be viewed, in at least many respects, as a different forensic person from the earlier Mr. Smith.

Of interest here is that the description of this case concludes with Mrs. Smith considering putting Mr. Smith in a nursing home, despite their mutual promise that one would care for the other at home even in the case of dementia. This offers an instructive contrast to Mrs. Jones's decision and reinforces the picture of Mr. Smith as a different forensic person at the beginning and end of his disease. The issues are not simple here. Mrs. Smith obviously still views Mr. Smith as her husband in the fundamental sense we discussed in the case of Mr. Jones. She views herself, that is, as in a unique position with respect to safeguarding his well-being. She sees herself also as having clearer obligations to respect the degree of autonomy and self-determination Mr. Smith is able to exhibit (despite their deviation from the wishes of the earlier Mr. Smith) than did Mrs. Jones, because Mr. Smith is capable of a higher degree of autonomy and self-determination than was Mr. Jones. Precisely because he is capable of leading something more like the life of a forensic person, however, and a life radically discontinuous with that of the earlier Mr. Smith, he seems more like a new forensic person than did Mr. Jones. For this reason (among others, of course), it seems more natural for Mrs. Smith to withdraw from the relationship, functioning independently and feeling that it may be legitimate for her to consider herself released from earlier promises. To the extent that Mr. Smith now is a different forensic person from the earlier Mr. Smith, the promises made to the earlier Mr. Smith are no longer in force.

This case, like the case of Mr. Jones, shows that forensic personhood is a matter of degree and that one can remain a forensic person in some respects but not others. It also shows that forensic personal *identity* is not an all-or-nothing affair. There are some purposes for which Mr. Smith should be treated as a single forensic person and others for which he should not. An account of forensic personal identity should be able to express this distinction. It should also take into account the importance of concern and affect in forensic personhood.

Mr. Garrison's case has two phases: first, his fall into apathy, related to Parkinson disease, and then the recovery of interest and personality change brought about by deep brain stimulation (DBS). The first phase, in which he develops apathy syndrome, is in many ways like the case of Mr. Smith, although Mr. Garrison's impairments, even at their worst, seem less profound than those sustained by Mr. Smith. The second phase, however, raises some new issues. While apathy syndrome diminished Mr. Garrison's forensic personhood, DBS increased it dramatically, almost to the level of his pre-apathy self. Taking the case as a whole and comparing it with that of Mr. Smith, we find that, in the end, Mr. Garrison retains a much higher degree of forensic personhood than does Mr. Smith and there is, correspondingly, an even greater inclination to say that (in at least some respects) he has become a different person. The post-DBS Mr. Garrison has interests, enthusiasms, and passions. He is able to make commitments to causes and to people and make important decisions. The passions and commitments that guide him after DBS, however, are very different from those that guided him before. It is thus easy to see him as a forensic person post-DBS, but hard to see him as the *same* forensic person as before DBS.

The issues are complicated, however. Despite the sense in which it seems we have a rupture of forensic identity in this case, no real question is ever raised as to whether Mr. Garrison should stay with his family after his treatment. Here, even more than in the first two cases, we have a strong sense that this is, of course, Mrs. Garrison's husband and he should retain his role in the family to whatever degree possible. Mrs. Smith's feeling that her husband was a forensic person to a fair degree, but not the one she married, led to the judgment that it was legitimate for her to think hard about whether she was bound to honor fundamental commitments made to the original Mr. Smith. There seems to be no corresponding sense that perhaps Mrs. Garrison should not honor fundamental commitments to Mr. Garrison. I see several reasons for this asymmetry. For one thing, Mr. Garrison, unlike Mr. Smith, is able to be an active participant in the continuation of the Garrisons' spousal commitments and is able to reciprocate. This is not important simply because Mrs. Garrison is "getting something" for her effort—we saw in the case of Mr. Jones that this is not the issue. But Mr. Garrison also understands himself as having a fundamental commitment to his wife and family. He is well enough to consider dissolving this commitment and does not do so. Because he is functioning well after his treatment, he is able to respond to claims made on him and adjust his behavior in ways that perpetuate and reinforce the earlier commitments. For this reason, even though he is changed, he is able to continue the commitment on his part in a variety of ways, and this helps reestablish his forensic identity. To the extent that Mr. Garrison is still participating in the family in this way, he is continuing to function as the same forensic person.

There are limits, however, on Mr. Garrison's capacity to keep his commitments, and this is where things get interesting. While he seems clearly to recognize that he should have a role in maintaining the household and interacting with his family, Mr. Garrison also enthusiastically and unilaterally follows his newfound passions and enthusiasms, violating explicit and implicit promises that he made to Mrs. Garrison. We can assume that the Garrisons had mutually agreed-on plans for their retirement and charitable giving. With Mr. Garrison's personality change, what had been agreed on is out the window. Now, if Mr. Garrison were simply the same forensic person, Mrs. Garrison would have every right to take him to task for this behavior. She could rightly complain that they had agreed on how their money would be spent and when they would retire and that he had violated these agreements without consulting her or against her wishes. But somehow, in the context of Mr. Garrison's history, these complaints—though they might well be made for pragmatic purposes—do not seem to bear the same moral weight that they would for someone who had not undergone DBS.

The relevant factor here is the *cause* of Mr. Garrison's change of heart. In any marriage, of course, people may change and rethink old promises in a way that alters old agreements. A Mr. Garrison counterpart (let's say, Mr. Harrison) might, for instance, have started talking to a new colleague at work who convinced him of the evils of the Republican Party or the need for more serious attention to environmental issues. Mr. Harrison might then start talking about these issues at the dinner table, laying out for Mrs. Harrison the arguments that had persuaded him and listening to her counterarguments. In the end, he might have talked her into changing their original plans. Failing that, the Harrisons might decide to split their resources in ways they had not before and to act more independently, or even decide that their differences had become too great to remain married. Any of these courses could be part of the natural history of a relationship. In Mr. Garrison's case, however, his change of heart is not caused by something he has seen, or learned, or reconsidered; it is the effect of direct electrical stimulation to the brain. He may certainly be able to present reasons to support his newfound views, and they may even be good reasons. But knowing his history, it is hard not to believe that their real origin is in his treatment. Mr. Garrison himself thus has a much different relationship to these commitments and the actions that flow from them than most of us do to our commitments and actions. To the extent that these new attitudes are the result of a medical intervention, it does not seem entirely fair to hold Mr. Garrison to the same standards of consistency that we would expect from someone who had not had such treatment. It does not seem obvious that he should be held responsible for commitments made before the personality change, and to this extent it seems right to treat him as a new forensic person.

The point runs even deeper, because it seems to affect not only Mr. Garrison's forensic personal identity but also his degree of forensic personhood. Someone who met Mr. Garrison for the first time after his treatment and did not know his history might well think he was as much a forensic person as anyone. However, knowing his history may well raise some questions about the degree to which he can be held responsible for his actions. In some sense, he clearly has the competence to make decisions and take responsibility for his actions. At the same time, however, it is not clear that his family has the right to complain about his self-centeredness and his failure to attend to what they say. Knowing that these traits appeared for the first time after his DBS makes us suspect that they are the direct result of brain manipulation. We are all, of course, ultimately at the mercy of our history and biology, to at least some extent, but the way in which Mr. Garrison's psychological makeup depends on biological factors is different from the ordinary case. How different depends on just how much control he actually has over his behavior, and it is not easy to determine this. What seems clear, however, is that the mechanism of change in Mr. Garrison's case raises important questions about both his degree of forensic personhood and his forensic identity.

Finally, there is the case of John Fast. This case recapitulates many of the features already discussed. During the period of steroid abuse, Mr. Fast's forensic personhood is compromised. While he retains a high degree of forensic personhood even on steroids, his ability to engage in the practices that define forensic persons is decreased. Because Mr. Fast's personality change while taking steroids is a result of chemical changes rather than ordinary personal evolution, the issues that arose in the case of Mr. Garrison's response to DBS arise in this case as well.[4] What is new here, however, is that in one telling of the story (ending 2), the changes to Mr. Fast are reversed, and he goes back to being someone who is not only undoubtedly a forensic person but also undoubtedly the *same* forensic person he was originally. The very fact that he is ashamed and embarrassed by his behavior while taking steroids —that his whole family may be embarrassed by it—shows that, for forensic purposes, this behavior is considered by everyone to be a part of *his* life. Steroid abuse is his mistake, and he is accepting the consequences.

The puzzle that arises here is that this same life segment is evaluated differently in the version of the story (ending 1) in which Mr. Fast does not quit steroids. In this telling of the story, the case is like that of Mr. Smith. Mr. Fast's decline leaves him with a greatly diminished degree of forensic personhood and is of a sort that makes it seem reasonable for his wife, friends, and teammates to view themselves as absolved from commitments made to the earlier John Fast. So in the second telling of the story (ending 2), Mr. Fast retains a higher degree of forensic identity during his period of steroid abuse than he does in the first (ending 1). In the second telling, his wife seems right to demand of him that he honor his commitment and get help; in the first, she seems right to tell him that she cannot be with him anymore. The problem is that the period of steroid abuse is exactly the same in both cases. It is not surprising, of course, that this case could have different trajectories and that there might be differences in its ultimate outcome. What is a bit harder to explain is the claim that Mr. Fast remained the same forensic person throughout the period of steroid abuse and recovery in the second case, but took on a new forensic identity in the first. An account of forensic personhood and personal identity should help make clear how the end of his story can affect his personhood at a stage that has already passed.

This survey of the four case studies thus reveals some aspects of forensic personhood and personal identity that an account of these concepts should explain, and it also raises some questions that we would like such an account to help us address. As we have seen, there are various degrees and facets of forensic personhood and personal identity, and affective as well as cognitive capacities are crucial to forensic personhood. We have seen that the causes of personality change bear on the issue of forensic personal identity; identity is compromised when change is brought about by disease or medical intervention in a way that it is not when the change is part of natural personal development. We have also seen that the ultimate reversibility of change can alter its effects on forensic personal identity. All of these elements should be part of an account of forensic personal identity. The questions raised by these cases mostly involve issues of how to assess the level of forensic personal identity and so determine the extent to which someone's current conduct should be judged in the light of past commitments and actions. I turn now to a sketch of my proposed account.

The Narrative Self-constitution View

The *narrative self-constitution view* of forensic identity, roughly put, holds that someone constitutes herself as a forensic person and creates her identity as such by forming an autobiographical narrative—a story of her life.[5] This means that being a forensic person involves coming to conceive of oneself as a forensic person, expecting one's life to follow the basic form of a forensic person's life. To do this, one must have a conception of where one comes from, where one is going, and how one's past, present, and future are interconnected. Certain normative expectations of how one's life should hang together are also necessary, as is responsiveness to demonstrations that one's life is not meeting these expectations. According to this view, one becomes a forensic person by having such a narrative, and only those with autobiographical self-narratives will count as forensic persons. The limits of a forensic person are set by the limits of a narrative. As long as a single narrative continues there is a single forensic person.

Thus far, this description is sketchy, and some detail must be filled in to make it comprehensible, let alone plausible. A full development of the narrative self-constitution view is obviously outside the scope of this discussion, but I will sketch some of its most basic features. First, it is important to be clear that forming and having an autobiographical narrative, in the sense meant here, is not something someone explicitly undertakes to do. The claim is not that to be a forensic person, one must write, speak, or even think anything like a formal autobiography. Rather, having an identity-constituting self-narrative is a way of processing experience, which involves (1) the general background assumption that one has a past and a future and that one's past, present, and future are interconnected in certain characteristic ways, and (2) an implicit knowledge of the basic details of one's own past and its connections to one's present and future. We do not typically experience what happens to us as completely without context. Experience comes to us as a basically coherent part of the ongoing story of our lives. When something *is* experienced as anomalous or without context, we typically feel compelled to address the anomaly, trying to understand how the dissonant event can be part of the life story we view as ours. To have a narrative in the relevant sense is thus to have a background understanding of oneself as the protagonist of an ongoing life story and to process one's experience in the context of that story. This last point is important. It is essential to realize that having a narrative is a matter not just of passively *knowing* one's story but of interpreting one's situation and deliberating in the light of it. Self-narration involves *shaping* one's life into a coherent story as well as *conceiving* of it as such. Understanding one's life as a coherent story is a means to *acting* in a way that will keep it coherent. Again, this is not to say that we must take this coherence as a conscious goal, only that this ideal plays a regulative role in the way we conduct our lives.

As far as we know, no one emerges into the world with a full-blown self-narrative, and probably no one *decides* to construct the kind of narrative I am describing here. The construction of a narrative self-conception is something that ordinarily functioning humans are socialized into in cultures that have and make use of the concept of a forensic person. This is, at least, the paradigm for the construction of an identity-constituting self-narrative. Children are taught that they are forensic persons by being taught that their lives and experiences should hang together in certain ways—that they should know what happened to them, learn from their mistakes, remember significant others, plan for the future, forge and maintain relationships, think about what they are going to be when they grow up and how they are going to become that, and so on. We learn from childhood that our lives are expected to have a general kind of shape and to cohere in certain ways. We learn also that we should be in a position to answer questions about how we came to be where we are, what we plan to do next, and why.

This is the basic framework of the narrative self-constitution view. As will be obvious on reflection, there are many aspects of this view that need development. Most especially, it is crucial to say something more about the shape that an identity-constituting narrative must have. I have said that one must expect "certain kinds of connections" between past, present, and future but have not said much about what those connections are. I cannot discuss fully all of the issues that need to be addressed to fill in these details, but some light might be shed by looking at two of the most important constraints on the kind of narrative that determines forensic identity. Failure to meet these constraints will compromise forensic personhood, forensic personal identity, or possibly both.

The first of these constraints I call the *articulation constraint*. Although self-narration is mostly implicit, to constitute a forensic identity a person needs to be able to articulate her narrative locally when it is appropriate to do so. This means that she should be able to answer questions of the following sort: "Where do you live?" "Are you married?" "Are your parents living?" "How did you come to live in California?" "What do you plan to do after you finish this project?" "Why did you buy that car instead of the one you told me you were going to buy?" In other words, a forensic person should be able to articulate both the basic features of her history and life situation—the facts of her autobiography—and the way in which her life hangs together, providing explanations for why she has acted as she has and why things have unfolded as they have. This does not mean that a forensic person can never fail to recall some detail of her past or that she can never act on impulse or whim. What it does mean is that she should not simply be at a loss about the basic facts of her autobiography or the internal and external elements that drive those actions and circumstances.

The articulation constraint requires not only that a forensic person should have *some* answer to questions about her history and motivations, but also that the answer must cohere with the rest of her narrative as revealed in other articulations and in her life. This constraint is required because forensic personhood involves not just suffering a history but participating in it and shaping it. Forensic persons are centrally agents, and agency requires not just understanding one's situation but acting on reasons. An agent is able to form plans and intentions and commit to courses of action. Agency also requires that one be able, at least in principle, to explain the unfolding of one's life to others. The narrative of a life will thus involve an interaction between the circumstances in which someone finds herself—an inner life that is, at least partially, developed in response to those circumstances—and deliberations, plans, and intentions that are developed within the context of a background awareness of both her external circumstances and her inner life. Someone who cannot produce (after some reflection) a coherent account of why she acts as she does will obviously be deficient as an agent with respect to those actions she cannot explain. When incoherence is pervasive in a narrative, the individual will be less fully a forensic person than someone who can articulate her narrative. When incoherence is local, the narrator may be a forensic person to a high degree, but the action or circumstance for which she cannot produce an account will not be integrated into her forensic identity.

The second constraint on identity is the *reality constraint.* This constraint requires that an identity-constituting self-narrative should fundamentally cohere with reality.[6] This is not to say that a narrative must be totally accurate in every regard or contain no trivial mistakes, but it should exhibit a fundamental grasp of what the world is like. It also implies that when a narrator is confronted with decisive evidence that some aspect of her self-narrative is factually inaccurate, she should alter it accordingly. The reality constraint can rule out both profound delusions and more minor misconceptions as genuinely identity-constituting aspects of a narrative. Someone who claims, for instance, that she is Napoleon and led the troops at Waterloo can be questioned about how tall she thinks she is and what sex she is, about how long she thinks humans typically live, when the battle of Waterloo took place, whether she is married to Josephine, and so on. Presumably, such a person will not be able to put together a coherent story of her life that both includes her having led the troops at Waterloo and maintains accuracy with respect to general facts about the world and specific facts about her own situation.

In a less extreme and more common sort of case, someone might remember having worked for one year at a particular job in 1985 and put this on her resume. If she finds evidence that the company for which she thought she worked did not exist before 1986 and employment records showing that she was hired elsewhere in that year, we would expect her either to revise her claim about when she worked for the company or to provide us with some reasonable explanation of why she does not accept the evidence of her employment records. The narrative self-constitution view is thus able to deny both the claims about Waterloo and the claims about employment in 1985 as actual parts of the narrator's history.[7] The motivation for this constraint is straightforward. Without a basic grasp of reality, one is not in a position to engage in the practices associated with forensic personhood.

There are, of course, many different degrees of narrative cohesion and many different degrees to which either of these constraints on an identity-constituting narrative can be met. The requirement that an account of forensic personhood should allow for degrees is thus easily met. The more fully developed and cohesive someone's narrative, the more fully a forensic person she is. To the extent that her narrative is compromised, so is her forensic personhood. This view can also easily account for degrees of forensic personal identity. A person's forensic identity is determined by her narrative, so when a narrative is disrupted or discontinuous, the degree of identity is correspondingly decreased. This view, as we shall see, can also meet the other criteria for an adequate account of forensic personhood laid out at the end of the last section and can help us address some of the questions raised by the four case studies.

Applying the Narrative Self-constitution View to the Four Case Studies

In discussing the four case studies, we uncovered several characteristics that an account of forensic personhood and personal identity should have. We have already seen that the narrative self-constitution view has the first of these: it allows for degrees of forensic personhood and personal identity. It is also fairly easy to see that this account has the other characteristics as well. From the cases of Mr. Smith and Mr. Garrison, for instance, we learned that forensic personhood depends on affect as well as cognitive capacities. The importance of affect is well-represented in the narrative self-constitution view. Autobiographical narratives are ways of knowing about not just our histories but our affective relations to them. If someone is deeply alienated from past actions or experiences, even if she can remember them well, they will not play their full role in providing a context for present and future experience and so will be included to a lesser degree in the person's narrative, yielding a lesser degree of forensic identity between the person who took the actions and had the experiences and the present person. When someone has a more pervasive affective deficit, like Mr. Smith or Mr. Garrison during their apathetic phase, the inability to have the usual affective relation to actions and experiences—to care about what happens to one and how one's life unfolds—compromises one's narrative capacity more generally and so decreases the degree of forensic personhood.

This analysis also shows how the narrative self-constitution view can account for the fact not only that forensic personhood and personal identity admit of degrees, but also that they have different facets or aspects, which can come apart. Someone may be able to narrate her life in the sense that she can accurately report on the course of her life but be unable to have the appropriate affective relationship to her actions and experiences; or her affect may be in place, but cognitive deficits may prevent her from accurately articulating what has happened. Either circumstance will diminish narrative capacity and influence the ability to engage in the practices of forensic personhood. These are, however, different ways in which narrative capacity can be compromised, and they have different implications for forensic personhood. This is, roughly, what the cases of Mr. Jones and Mr. Smith demonstrate.

The case studies also reveal that the mechanism of personality change is important to its effect on forensic personhood and identity. The changes to Mr. Garrison's personality were disruptive to his forensic personhood and identity in a way that natural personal development would not have been, because they were caused by deep brain stimulation. The narrative self-constitution view can explain this as well. Seeing oneself as the protagonist of an identity-constituting narrative requires that one consider oneself as an agent, acting for reasons. The articulation constraint, you will recall, requires that a person should be able to explain why she does what she does. It is essential that one see one's actions as flowing from one's plans, projects, intentions, beliefs, and desires, if these are to be the actions of a forensic person. In the case of Mr. Garrison and others whose actions are significantly affected by direct manipulation, there is a tension in meeting this demand. If Mr. Garrison articulates his history without running afoul of the reality constraint, he will have to acknowledge that his current passions and interests—the things he takes as reasons—were caused by manipulation of his brain. It is not because he learned something important about the defects of his previous commitments that he gave them up; it is because he had electrodes implanted in his brain. Having acknowledged this, however, Mr. Garrison will have difficulty taking his current commitments seriously as reasons for action. Knowing that his enthusiasm for the Democrats or the environment is brought about by direct stimulation of his brain should make him suspicious of these passions as genuinely reason-giving.

Mr. Garrison is thus in something of a bind with respect to his self-narration. If he is to consider himself as an agent, take his current reasons seriously, and act accordingly, he will need to start anew—with the passions he finds himself with after DBS and not questioning too hard how he came to have them. This will impede his ability to fully meet the articulation constraint. If he does articulate the source of these passions realistically, he will either have to feel his agency compromised by their anomalous origins or will have to ignore them and act on his previous commitments, even though he is alienated from them. In either of these cases, his narrative will be compromised because he does not fully consider himself an agent. Alternatively, Mr. Garrison can come up with a story about how his newfound enthusiasms are really the result of personal development rather than DBS, but this would make him run afoul of the reality constraint. When there is a change in personal commitments caused by illness, medical treatment, accident, or brainwashing, there is no narrative solution that does not result in a decreased degree of narrative coherence and so a decreased degree of forensic personhood and/or identity. The best solution seems to be the first—taking the newfound enthusiasms as a starting point and critically evaluating them from where one stands now, without much attention to one's earlier views and commitments. This will maximize one's degree of forensic personhood but decrease forensic identity with one's earlier self.

Finally, the narrative self-constitution view needs to account for the fact that our judgment of the degree of forensic personal identity between the original Mr. Fast and the steroid-taking Mr. Fast depends on how the story turns out. To see how it does this, we must remember that narratives unfold over time and that we cannot always tell at once whether a nonstandard narrative will work to constitute the life of a person. People are complicated, and their narratives are going to be as well. As with most complicated stories, there will be times when the direction in which the story of a forensic person's life is unfolding is unclear even to himself—sometimes, unclear to the point where it is not obvious that there is a *story* at all. This phenomenon is familiar from nonstandard narrative in fiction or film—the movie *Memento*, for instance, or the nonlinear films of Quentin Tarantino. The viewer is likely to be confused for a good portion of these films, but as time goes by and more information emerges, most people are able to start sorting the events into a coherent narrative. The nonlinearity of the presentation is somewhat disconcerting, but in the end it yields a comprehensible story. This is not true for every experimental film or work of fiction. Sometimes things never gel enough to create a satisfying narrative. The problem is that it is not always obvious at the outset whether a coherent story will emerge eventually or not. Thus it is often worth sticking with a confusing presentation for at least a little while to see whether it ultimately comes into focus.

Something analogous can happen with a person's self-narrative. The narrative self-constitution view demands that we form a coherent self-narrative and that we understand the parts of our life histories to coalesce in certain ways. This does not mean that we can never be perplexed about ourselves or fail to understand, for the moment, where we are going. When this happens, we—and our loved ones—need to exhibit a bit of patience to see whether the meanings that escape us will become clearer and whether the present will ultimately become comprehensible as part of a coherent life narrative. We need to try to understand ourselves when there is narrative dissonance, but we also need to be able to tolerate such dissonance, at least temporarily.

The two versions of Mr. Fast's case demonstrate the two possible outcomes of such tolerance. In the story with ending 1, in which he continues his steroid use, it becomes clear that his narrative has been disrupted in a way that causes a deep break between past and present. It will, of course, be possible to put together a factual history of Mr. Fast's life that makes a certain amount of sense, but a fully coherent narrative—that is, one that is coherent with respect not just to facts but also to motives—will be no more forthcoming in this case than in the cases of Mr. Smith and Mr. Garrison. With ending 2, however, when Mr. Fast returns to his original personality, he, himself, is able to tell a story in which he appropriates the period of his psychosis. The story might involve too great a desire to excel as an athlete, or insecurities or external pressures that led him to make a bad decision. It will also include the factors that led him to recovery. As noted earlier, the fact that the recovered Mr. Fast feels ashamed and embarrassed by his actions while using steroids shows that he takes that portion of his life to be part of his life. Even in this second version of the case, the disruption caused by his period of psychosis makes Mr. Fast's self-narrative less fully coherent than it would have been without such an episode, but it is far more coherent than the case in which he does not recover. It is a characteristic of narratives that the significance of any event in the narrative depends on its role in the whole story. In the second version of Mr. Fast's case, his recovery makes the period of steroid abuse a coherent part of a single unfolding life, albeit a part in which he was a forensic person to a lesser degree than in other parts. In the first version, this does not happen.

The narrative self-constitution view is thus able to explain the elements of forensic personhood and personal identity we observed in thinking about the four cases. I next discuss how this view can be applied to some of the questions raised by these cases.

Addressing Questions

The questions about forensic personhood and personal identity that grew out of our discussion of the four case studies are fundamentally practical questions. In each case, the patient's wife faces a variety of dilemmas about how she should treat her husband throughout the course of his illness. The narrative self-constitution view cannot resolve these dilemmas by giving concrete answers to questions about what these women should do. We should be suspicious of it if it did. These situations are enormously complex, as are human relationships, and the right thing to do in each case will depend on a great many factors, some of which—such as the availability of financial resources and viable options—have little to do with personhood or personal identity. What the narrative self-constitution view can do, however, is help us get clearer on what is at issue in these dilemmas, and this clarity can be an invaluable aid in deciding what to do.

To enter into the details of each question raised here and describe specifically how the narrative self-constitution view would apply requires a far too extended discussion. Instead, I will speak to some of the basic principles at work. The questions that arise about how to treat the men who are the subjects of these case studies are, for the most part, related to the value we place on forensic personhood. I said earlier that we take forensic personhood as a good and believe that we should, as far as possible, respect and nurture the forensic personhood of others. Forensic personhood is not, as I have noted, the only or even necessarily the highest good. It is, however, an important part of human interaction, and so the obligation to respect and support forensic personhood in oneself and others is a serious one. The dilemmas that arise in these case studies stem from circumstances in which it becomes unclear how this principle should be pursued.

The problem of honoring forensic personhood in these cases tends to take one of three forms. First, there can be uncertainty about the degree to which someone remains a forensic person, and so uncertainty about the degree to which it is appropriate to treat him as such. While failing to acknowledge forensic capacities in those who possess them seems injurious, it seems cruel to expect those capacities in those who do not possess them. Mrs. Garrison, for example, may be puzzled about whether she should hold Mr. Garrison responsible for his self-centered behavior or ignore it on the grounds that it is the result of his treatment and he cannot help it. Mrs. Smith may have similar worries about the extent to which she should express expectations of appropriate behavior from Mr. Smith. Second, there can be questions about which action will most respect whatever degree of personhood is present. Mrs. Jones's dilemma about whether to honor the wishes of the past or present Mr. Jones when these are in conflict is a good example of this sort of difficulty. Finally, there is the question of how best to nurture forensic personhood, helping the affected individual maintain the highest degree of personhood that he can. We can imagine Mrs. Fast, for instance, wondering whether leaving her husband or staying with him is more likely to help him stop abusing steroids.

These three types of question are interrelated in ways that the narrative self-constitution view can help us clarify. This view holds that one's self-narrative, and so one's forensic personhood, is something one *constitutes.* One is not born a forensic person, one makes oneself a forensic person by forming a narrative self-conception. But we do not form self-narratives in a vacuum. As I pointed out earlier, children are *taught* to narrate their lives. This happens in a variety of ways. They are reminded of what they have done and prompted to anticipate what they are going to do. We talk to them about plans and give them age-appropriate responsibilities and freedoms. We let them suffer the consequences of actions and help them consider how to avoid those consequences next time. This is not, of course, something we do as a separate activity in "narrative school"; it is the way we interact with children. Gradually we treat them more and more like forensic persons, and this helps them develop the abilities that make them forensic persons. After these abilities are fully developed, the narrative capacities of adults are supported and maintained as others treat them as fully forensic persons. It is by holding others responsible for what they do and respecting their rights of self-determination that we reinforce their narrative understanding of themselves and hence their forensic personhood.

In ordinary circumstances, then, we nurture and support forensic personhood by respecting it. The degree to which someone is a forensic person is not totally independent of the degree to which he is treated as one. By systematically failing to treat an ordinary human being as a forensic person we might well greatly decrease his degree of forensic personhood, and by systematically treating him as a forensic person we are likely to greatly increase the degree of his personhood. There are, however, obvious limits. Treating someone as a forensic person involves having certain sorts of expectations and making certain sorts of demands. For this attitude to increase someone's degree of forensic personhood, he must have the fundamental capacities necessary to meet those demands. If he does not, treating him like a forensic person will at best have no effect on his degree of forensic personhood and may well decrease it.

We see this in the case of children through the importance of making age-appropriate demands. Talking to your seven-year-old about what he is going to be when he grows up will help establish the basic contours of a narrative self-conception. Pressing him mercilessly on the details of his plans for becoming an astronaut or grilling him about why he changed his mind about being a firefighter is another matter. If we demand from him the consistency we would expect from an adult in considering employment opportunities, we are not going to enhance his ability to think of his life as a narrative; we are going to confuse him. Far from leading to a more coherent self-understanding, such treatment is likely to undermine what narrative sense exists.

This same principle applies in the case histories we have been discussing. There we encounter adults who have lost some of their capacity for self-narration and for forensic personhood. Respecting and nurturing their personhood, diminished though it may be, is an important aspect of treating them well. Yet it is crucial here, too, that the level of expectation be realistic. If it is not, it is likely to be undermining not only to the patient's happiness but also to his forensic personhood. This is why, in each case, such delicate questions arise about what kinds of demands are to be made. What is at issue is figuring out which expectations, given the existing level of forensic personhood, will enhance the degree of personhood and which will be undermining. This is a large part of what is happening when Mrs. Jones wonders whether she should in any way hold the Mr. Jones with severe dementia to the standards of conduct of the previous Mr. Jones; when Mrs. Smith wonders how to deal with Mr. Smith's inappropriate behavior; when Mrs. Garrison does not know how angry to get about the diversion of funds; and when Mrs. Fast does not know whether to stay or go.

The narrative self-constitution view thus helps us see that, in wondering to what extent someone is appropriately treated as a forensic person, we are asking about the extent to which he is still able to form a coherent self-narrative. This kind of question can be addressed by determining to what extent narrative capacities have been compromised and in what ways. In wondering how best to respect someone's forensic personhood, we are registering the fact that sometimes respecting someone's forensic personhood by demanding consistency over time—one of the ways in which we normally treat someone as a forensic person—can interfere with our ability to treat him as a forensic person in the present. Finally, in wondering how best to nurture and encourage someone's forensic personhood, we are wondering how best to help him develop and maintain the narrative capacities he has.

Seeing the questions raised by these cases as questions about narrative does not yield easy answers, but it does give us a fruitful way to think about them. Further analysis of the different capacities that go into self-narration and the way in which they support forensic personhood will give concrete aid in attempts to sort out some of these difficult issues. I offer some analysis along these lines in *The Constitution of Selves* (Schechtman 1996), and there is reason to hope that work in empirical psychology can flesh out these ideas more thoroughly. No amount of fleshing out is likely to provide definitive answers about what to do in these cases, but it could provide useful insights. Issues of forensic personhood and personal identity are complicated in ordinary circumstances, and in unusual ones they become almost impossibly so. It is, however, precisely because these issues are so important to us and play such an important role in our lives that we need to understand them as well as we can. In our everyday life, we will obviously continue to interact with others on many levels and in many ways, of which the interactions specific to forensic personhood will constitute only one part. Even for the interactions specific to forensic personhood, we will probably continue to reidentify people by reidentifying bodies and to presume that all human beings are forensic persons. When these assumptions break down, however, we need to have tools to help us determine what to do. The narrative self-constitution view, I hope, can provide one resource in the daunting but exciting task of figuring out how to treat one another.

NOTES

1. Dissociative identity disorder is also known as multiple personality disorder.

2. Locke thinks that his view speaks to metaphysical questions about personhood and personal identity and so sees somewhat broader application for his theory of identity than those discussed here. While I think that the connection between issues of forensic identity and broader metaphysical identity questions is an important one and is central to understanding philosophical issues of personal identity, discussing it would take us too far afield here.

3. At this point, a legitimate question is who the *we* who value forensic person-

hood are. What I am describing here, and will rely on later, is basically the fact that modern Western culture values autonomy and takes it to be a good. It is not clear that this value is universal. Historically there have been cultures that do not place the same value on autonomy (or understand "autonomy" in the same way) that modern Western culture does, and arguably this is also true for some currently existing cultures. For the purposes of the present discussion, the claim that forensic personhood is a good should be taken as more descriptive than normative. The question of whether our culture *should* value these capacities to the extent that it does is obviously deeply important, and just as obviously is outside the scope of this chapter. Cases of the type we are considering raise particular kinds of question for those in a culture that values forensic personhood as we do. My interest here is in understanding how these questions can be addressed, given a background assumption of the value of forensic personhood. This discussion might, and probably should, provide both an occasion and some resources for asking more fundamental questions about the value of forensic personhood, but that is a matter for a separate and much larger inquiry.

4. In both cases, of course, the changes observed stem indirectly from a volitional choice—in Mr. Garrison's case to undergo DBS and in Mr. Smith's case to take steroids. Once these decisions are made, however, the results follow in a causal way, and while the voluntariness of the initial action might well affect our degree of sympathy for those who took them (especially in the case of Mr. Fast), it does not say much about how much volitional control there is after the fact.

5. I offer a full development of this view in *The Constitution of Selves* (Schechtman 1996). In that context, I intended the view to have a somewhat broader application than the one I describe here. However, because I believe that these broader applications are intimately connected with questions of forensic personal identity, the view is developed here to serve as an account of forensic identity.

6. I do not insist that we have some uncontroversial way of determining the nature of reality. Here I am employing a commonsense, everyday notion of reality rather than a metaphysically loaded one.

7. This constraint may seem to be just a way of sneaking back in a bodily account of personal identity. How, after all, are we to tell whether the person claiming to be Napoleon is not really Napoleon if we do not rely on the evidence that she obviously does not have Napoleon's body? The reality constraint often depends on observing facts about the body. It does not, however, turn the narrative self-constitution view into a bodily account of identity. For one thing, this view allows that it is at least logically possible that this person is the same forensic person as Napoleon. If she can give us a convincing account of how she came to be in a new body—an account that, for instance, revealed some kind of secret technology of which we were unaware—she might convince us that she is Napoleon. The narrative self-constitution view, more generally, allows for the possibility of more than one forensic person inhabiting the same body. Finally, this view does not make the continuity of a single human being a *sufficient* condition for forensic personal identity, as would a view based on human identity.

CHAPTER FIVE

Personal Identity and Choice

Carol Rovane, Ph.D.

It was John Locke who formulated the problem of personal identity as it is understood by contemporary philosophers (Locke 1975). He asked whether personal identity is the same as human identity, and he answered no, on the basis of the following interrelated considerations. First, he offered a definition of the person as "a thinking, intelligent being that has reason and reflection and can think itself as itself in different times and places" (335). He went on to note that persons are aware of themselves and their thoughts through consciousness. He concluded that the life of a person extends exactly as far as the person's *consciousness* extends—by which he meant the person's *reflexive awareness of its existence over time through memory.* It would seem to follow directly from these considerations that the life of a person is not to be equated with the life of a human being. The life of a person, in Locke's sense, would seem to begin not with biological birth but, rather, with the dawning of self-consciousness, which comes much later in human development. Likewise, it would seem that the life of a person may also end before biological death, if personal consciousness is lost sooner than that, either through amnesia or through a total loss of consciousness. Locke also raised the possibility of something like dissociative identity disorder, with an imaginary case in which two distinct consciousnesses—a "day person" and a "night person"—alternate within the same human being. But the possibility that held the greatest interest for him was much farther removed from the actual human condition. It is the purely hypothetical possibility that he raised with his famous thought experiment about a prince and a cobbler. The thought experiment asks us to imagine that the consciousnesses of a prince and a cobbler are switched, each into the other's body, and to work out who would be who after the switch. Locke took it to be obvious that the prince would be the person with the princely consciousness and the cobbling body and the cobbler would be the person with the cobbling consciousness and the princely body. And he took this to be a decisive ground for concluding that personal identity and human identity are two different things.

In my view, this last possibility—of "body-switching"—lies too far beyond the actual facts of the human condition to be of interest to anyone but writers of science fiction. But many philosophers do not share my dismissive attitude. They insist that it is instructive to think through how our concept of a person would apply in imaginary cases like the case of the prince and the cobbler. As a result, an entire literature of thought experiments about personal identity has arisen in which we are asked to give intuitive responses about who would be who in various counterfactual conditions in which persons allegedly survive in new and different bodies, such as brain transplants, brain reduplication, brain reprogramming, and so-called brain zippering procedures.[1] One difficulty with relying on such thought experiments is inconsistency of response: not everyone finds it intuitively plausible that the same person could persist in a new and different body. But even if everyone's intuitions accorded with Locke's, it would still be precipitate to conclude that he was right to draw a distinction between personal and human identity. Although we can easily *conceive* the possibility of body-switching, future science might eventually tell us that it isn't *really* possible because it is ruled out by the laws of nature —much as physics tells us that we can't fly like Peter Pan even though we find that easy enough to imagine, too.

If the entire case for Locke's distinction between personal and human identity rested on his thought experiment about the prince and the cobbler (and other similar thought experiments), then I would recommend that philosophers leave it to science to settle the matter. But as it happens, there are interesting grounds on which to affirm a *version* of his distinction that are *not* hostage to the deliverances of future science.

To uncover these grounds, we need to explore a distinction that Locke himself glossed over in his own account of personhood and personal identity. This is the distinction between *consciousness* and *rationality*. Consciousness concerns the phenomenological aspects of mental life, which philosophers often characterize in terms of *what it is like* for a subject to be in various conscious states, whereas rationality is an ability that reflective agents bring to bear when they *deliberate* about the various reasons they have on which to act.

Although Locke made consciousness central to his account of personal identity, there is no doubt that he conceived persons as rational agents. Indeed, for him, much of the philosophical interest of the problem of personal identity lay in certain ethical dimensions of personhood that follow on rational agency—most especially, personal accountability. What he primarily wanted from a philosophical account of personal identity was an account of what it would take for him to stand before God on Judgment Day and be held to account, as one and the same person, for all of his earthly actions. In this connection, he clarified that the term *person* is "a forensic term, appropriating actions and their merit; and so belongs only to intelligent agents capable of a law, and happiness and misery" (Locke 1975, 346). In the first section of this chapter, I briefly catalogue some of the other ethical dimensions of personhood that also follow on rational agency. And I will argue that these constitute important additional grounds for equating the concept of a person with the concept of a rational agent.

I said that Locke glossed over the distinction between consciousness and rationality. I should add that it is completely understandable that he did so. It is natural to assume that the activities of rational agents are conscious processes. And it would seem to follow fairly directly from this natural assumption that if persons are rational agents, then the condition of personal identity will coincide with the condition in which there is a single, abiding center of consciousness, just as Locke held. Most of Locke's opponents agree. That is, they do not deny that a person is a rational agent with its own, separate center of consciousness. What they deny is Locke's further claim that a given person's consciousness could be transferred to a new and different body. I've already made clear that I don't think philosophers should try to evaluate this further claim of Locke's from the vantage point of their armchairs; rather, they should leave it to future cognitive science to figure out what the nature and basis of consciousness is. But the same does not hold for the underlying assumption from which Locke's whole argument proceeded, according to which each rational agent has a separate center of consciousness.

I've conceded that Locke's assumption is a natural one—indeed, so natural that it is shared by most of his opponents. But all the same, I will argue that it is mistaken.[2] To be more precise, I will argue for the following claim: the sort of unity of consciousness that Locke took to constitute personal identity is neither necessary nor sufficient for being an individual rational agent. It is not necessary, because a *group* of human beings can function together as a single agent even though those human beings have separate consciousnesses; and it is not sufficient, because a single human being can house *multiple* rational agents, each of which is able to function independent of the others, even though their human host has a single, unified consciousness.

These two possibilities—of *group persons* who are larger than human size and *multiple persons* who are smaller than human size—provide us with a novel interpretation of Locke's distinction between personal and human identity; and this version of the distinction can stand even if his own version of it, in terms of consciousness, is eventually ruled out by future science. What is more, unlike the possibility (of body-switching) that he asked us to envisage in his thought experiment about the prince and the cobbler, these two possibilities can be established right now, without waiting for any deliverances from future cognitive science. All we need to appeal to is the self-understanding that we rational agents already possess, simply to function as the sorts of agent we are.

My claim that there can be group and multiple persons is controversial. It might also appear to be remote from the concerns in this volume. The possibility of group persons might appear to be especially remote from these concerns—remote both from the particular task of coping with the four case studies we have been asked to think about and from more general issues in biomedical ethics. But that appearance is misleading. A proper understanding of the group case is crucial for understanding my main thesis about personal identity, which is this: *the existence of a person is never a metaphysical or a biological given but is always bound up with the exercise of effort and will*. It is relatively easy to see why this should be so in the group case. Assuming that there could be such a thing as a group person, it is easy to see that such a thing wouldn't be a metaphysical or a biological given—a group of human beings couldn't possibly function as a single, unified agent without effort and will. But I'll be arguing that the same holds for all persons, including persons of human size. They, too, must exercise effort and will to function as single, unified agents—and the reason is that in remaining one, they are choosing against alternatives that are available, such as integrating into larger group agents or fragmenting into smaller, multiple agents. Understanding this will prove especially relevant to cases 3 and 4. It will also facilitate a broader understanding of how various medical and pharmacological interventions may affect personal identity. If I'm right, the role of personal choice in personal identity is not something that doctors, patients, and families can reasonably overlook or discount when they contemplate certain interventions.

Personhood

Let me begin by elaborating what I take to be the most important strand in Locke's definition of the person as "a thinking, intelligent being with reason and reflection." The things that satisfy this definition possess what I have been calling *rational agency.* Not only can they move about in the world and make a causal difference to what happens in it, but they can also deliberate about the reasons they have on which to act. And their common rational nature has a social significance as well. Rational agents can aim to influence one another by presenting one another with *reasons* concerning what they should think and do. This is something they do whenever they engage in conversation, argument, and criticism. And this brings to light one intuitively powerful reason that we should reserve the term *person* for rational agents: these forms of engagement are *distinctively interpersonal* forms of engagement. From here on, I shall take the capacity for such interpersonal forms of engagement as a criterion of personhood. And for reasons that will soon be clear, I'm going to refer to it as the *ethical criterion of personhood.*[3]

It seems to me that all persons implicitly grasp and employ this criterion of personhood in their everyday lives, even if they don't always express their understanding of it with the language of personhood. That is, we persons all understand that there are some things that we can engage in distinctively interpersonal ways, such as conversation, argument, and criticism, and other things that we can't engage in these ways—for example, our pets. We can love our pets. We can need them. We can lose games to them. We can even lose battles of will to them. But we can't lose arguments to them. And we know that. Thus we know that we have a nature that we don't share with our pets but do share with those whom we can engage in conversation, argument, and criticism. It is a nature that all rational agents possess, and we all recognize it as something that we all have in common—which I am proposing to track with the term *person.*

There is no gainsaying that this criterion of personhood is highly restrictive. It obviously disqualifies fetuses from having the status of a person and, even more controversially, also infants, young children, and all severely damaged human beings who have lost the capacity for distinctively interpersonal forms of engagement. However, we can accept the restrictive character of the criterion with equanimity if we bear in mind two crucial points.

First, the criterion is, in one important respect, highly inclusive. It says that whenever we engage others in conversation, argument, and criticism, we thereby acknowledge their personhood, at least implicitly; and it follows that any subsequent attempt to explicitly deny their personhood will necessarily be hypocritical. I believe that this has been true of many historically salient denials of personhood—for example, the personhood of African slaves in the American South and of women in various places and times. And it is a real virtue of the criterion that it exposes the hypocrisy of these and other attempts to exclude from the class of persons those whom we can engage in distinctively interpersonal ways. Indeed, this is one reason that I call it an *ethical* criterion, because it is ethically important to acknowledge the personhood of all those with whom we stand in distinctively interpersonal relations.

Second, the restrictiveness of the ethical criterion concerns only what should be counted as a person. It neither states nor implies that persons are the only appropriate objects of ethical concern. In general, ethical issues arise with the recognition that there are other points of view from which things matter, for then the question arises concerning what we should do in the light of the fact that there are other points of view from which things matter. And although persons are the only things that can raise and consider ethical questions, they are not the only objects of ethical concern, because they are not the only things that have points of view from which things matter. Many nonpersons do, as well. In fact, I've already mentioned some of them: our pets and other animals, infants, elders with senility, and so on. I'll not comment on the case of fetuses, which I leave for others to work out. I'll just say a bit more about the underlying issue.

I insisted in my introductory remarks that we ought not to gloss over, as

Locke did, the distinction between consciousness and rationality. My main point there was to draw attention to an assumption that he actually shared with his opponents but that I will be arguing against, namely, that each rational agent has its own center of consciousness—for that assumption is directly challenged by my argument for group and multiple persons. But here I want to emphasize a quite different reason that we need to take due account of the distinction between consciousness and rationality. Many conscious beings are not rational beings in the sense required for personhood. Yet things can go better or worse for them because there is something it is like for them to have the sort of consciousness they have—in other words, it *matters* to them what having the sort of consciousness they have is like. This makes them appropriate objects of ethical concern, even though they cannot themselves *raise* ethical questions in the way that persons can. So when we acknowledge the ethical importance of the category of *person,* we are not thereby restricting the domain of ethical significance to persons. In fact, the ethical criterion of personhood that I'm advocating leaves us free to extend the range of our ethical concerns as far as we like—not only to all conscious beings but also to all living things and, for that matter, to literally everything. I think it would be going a bit far to extend the range of our ethical concerns beyond things that have points of view. So, for me, the ethical question raised by fetuses turns on the question of whether they already have points of view. But it is worth registering that the ethical criterion of personhood doesn't preclude an even more liberal—you might say *ecological*—conception of the range of appropriate ethical concern, in which even mere things that have no points of view merit some form of ethical concern.

Another point worth registering is that the ethical criterion of personhood leaves us free to extend *rights* to nonpersons. This is a point to which I'll return in discussing the four cases, in the last section of the chapter. For now, I'll simply observe that there are many different sorts of rights and they have quite different grounds. Some rights are properly conceived as *personal* rights, in the sense that they can coherently be accorded only to those human beings (and other things) that satisfy the ethical criterion of personhood. This is true of most of the political rights associated with the liberal tradition, such as freedom of speech and conscience. These rights are special cases of a more general right of self-determination—that is, the right to live by one's own choices without undue influence from others; and the only things that could possibly have a right to such self-determination are rational agents who can *claim* it for themselves—or, in other words, persons in the sense tracked by the ethical criterion of personhood. In contrast, some of the rights that we now call *human* rights can coherently be accorded to human beings who don't satisfy the ethical criterion and who can't claim these rights for themselves. These would include rights to adequate nutrition, security, healthcare, and so forth. Of particular relevance to biomedical ethics, the ethical criterion leaves us free to extend the *right to life* to all human beings, including fetuses. For that matter, it leaves us free to acknowledge certain animal rights as well.

Why, it might be asked, should we be interested in such a restrictive criterion of personhood that disqualifies so many human beings from the class of persons? And why should we call it an *ethical* criterion, given that it doesn't track all of the appropriate objects of our ethical concern, or even all of the things that could be said to possess rights? *We should do so because even if the range of our ethical concerns can and should extend beyond the class of persons, and even if the same is true of certain human and animal rights, it remains true that the class of persons stands out as ethically distinctive.* There are many things to say about the distinctive ethical significance of persons, many more than I can say in this chapter or in any single essay. In my introduction, I observed that Locke singled out the fact that persons alone can be held to account for their actions. In this section, I have explained that they alone can acknowledge and treat one another *as persons* and recognize the prejudice that lies in denying one another's personhood. I've also explained that persons alone face ethical choices as well as occasion ethical concern. And I've noted that some rights can coherently be accorded only to persons. Let me now add: persons alone can exercise prudential self-concern; they alone can make and keep promises to one another; they alone can treat others with respect. Last but not least—in fact, paramount for the purpose of addressing the four cases—persons alone are *competent* in the sense at work in medicine and the law.

In concluding this section, I want to reiterate that there should be no worry that the ethical criterion of personhood is somehow prejudicial against fetuses and other helpless or legally incompetent human beings. It is a plain and undeniable fact about these human beings that they cannot engage in distinctively interpersonal relations or in any of the ethically significant activities I just catalogued of which persons alone are capable. We can acknowledge this fact and still place a moral premium on all human life in just the way that opponents of abortion and euthanasia do; we can accord certain rights to all human beings as well, and perhaps to animals; and we can extend the range of moral concerns even farther than that. All of this is compatible with acknowledging that there is a distinctive ethical significance that persons alone possess, by virtue of their being rational agents who are capable of engaging one another in distinctively interpersonal ways—which is what I aim to track with the ethical criterion of personhood.

Personal Identity

I'm now ready to explain what the ethical criterion of personhood implies about the condition of personal identity—more specifically, why it ushers in the novel interpretation of Locke's distinction that I briefly sketched in my introductory remarks, according to which there can be group persons composed of many human beings and multiple persons within a single human being.

This version of Locke's distinction doesn't follow *directly* from the ethical criterion. All that directly follows is that human beings do not qualify as persons at every moment of their biological lives—more specifically, that they are not persons when they are conceived or even when they are born, but only later, when their rational capacities are sufficiently developed. And we needn't infer that personal identity is distinct from human identity, in the sense that there is a distinct thing—the person—whose life is shorter than a given human being's life. We can suppose instead that personhood is a *status* that is sometimes achieved by a given human being and sometimes not, without introducing any distinct existence.

Clearly, it will take additional argument to move from the ethical criterion of personhood to any version of Locke's distinction. So let me now offer such an argument. My first step is to provide a slightly more detailed account of what it is to be a rational agent who is capable of distinctively interpersonal forms of engagement.

At the heart of this account is a simple point that is liable to be misunderstood: *rational agents are things that can grasp and respond to the normative requirements of rationality.* The reason this might be misunderstood is that it sounds like an overly idealized conception of what personal agency involves. So let me be clear. I am not suggesting that persons are perfectly rational. I am not even portraying persons as closely approximating the ideals of rationality. The only agents who could possibly satisfy the normative requirements of rationality would be God and angels. We mortals are bound to fall woefully short of meeting them. After all, it is impossible to be perfectly rational without working out all that rationally follows from our beliefs and other attitudes, and we are finite beings who do not have the time or the mental capacity to work all that out. In addition to this obvious fact, that our finite nature makes it impossible for us to be perfectly rational, we also suffer from various lapses that prevent us from being as fully rational as it is within our power to be—as when we are tired, careless, distracted, drugged, and so on. We are also prone to positive breaches of rationality, such as self-deception and weakness of will. In other words, we are prone to irrationality as well as to failures of full rationality. But none of our failures and breaches of rationality disqualify us from the status of *rational agent*. What makes us rational agents is not perfect conformity to the normative requirements of rationality; what makes us rational agents is that we can grasp those requirements and respond to them. And there are many forms that such responsiveness can take besides being perfectly rational. Perhaps the two most important forms are evaluative and ameliorative. Rational agents can discern when they fall short of meeting the requirements of rationality, and they can regard this as grounds for self-criticism and self-improvement. And this is all it takes to qualify as a rational agent —not the ability to be perfectly rational (which no actual agent has) but the ability to identify failures and breaches of rationality along with the ability to respond appropriately with self-criticism and efforts at self-improvement.

These remarks about the nature of rational agency carry over to what I've been calling distinctively interpersonal forms of engagement, for all such engagement aims at rational response, by which I mean an exercise of rational agency. Here, too, there may be misunderstanding. I may seem to be suggesting that persons are ideally rational beings who manage, in general, to influence one another by getting one another to have ideally rational responses to whatever they might say and do to one another. But my claim is far weaker and more realistic. I am pointing to a profound difference between two ways in which one person can aim to influence another. One way is merely physical (e.g., pushing, carrying, drugging). Another way involves communication, usually in the form of speech, but sometimes gesture will do (e.g., exhorting, advising, criticizing, protesting, bribing, threatening). When I communicate with another, my interlocutor cannot assume that I am being fully or ideally rational, nor can I assume that my interlocutor's response to what I say will be fully or ideally rational (as I just explained, this is impossible for finite beings). But, all the same, I am operating on a normative level. I am offering up to my interlocutor a consideration that she must take up and evaluate from her own point of view. If the consideration I offer up is faulty or ill-considered, it is still something that can be rationally evaluated. And the same holds for any response that my interlocutor might make. However faulty or ill-considered that response might be, it is still something that can be rationally evaluated. And this suffices to qualify the response as rational in the sense I am elucidating. Think of a particularly unimpressive debate between two adolescents in which both make claims for which they have inadequate evidence and both draw conclusions that don't follow anyway. However far short of perfect rationality two such adolescents might be, they are still attempting to engage in a rational activity. They are not merely pushing up against one another in physical space. They are interacting in a way that makes it coherent and appropriate to evaluate what they are doing in the light of the normative requirements of rationality. They are interacting in a normative space that some philosophers call the "space of reasons." This space includes all activity that can coherently be evaluated in the light of the normative requirements of rationality. Even our most helpless and most inescapable syndromes of irrationality take place within this space. Otherwise they could not be called irrational—they would be mere happenings that, like the weather, are not subject to rational evaluation. The whole point of the label *irrational* is to remind us of the contrast between what we are actually doing and what it would have been rational to do instead.[4]

I want to say a bit about what the normative requirements of rationality are. I won't try to give an exhaustive specification of them, but just a few examples to bring out something they all have in common. My characterization of these requirements will again be highly idealized. But this should not be construed as departing in any way from what I've just said about the ways in which persons typically fail to be fully or ideally rational.

The most general normative requirement that rationality imposes on a person is that the person should arrive at and act on all-things-considered judgments about what it would be best to do in the light of all her beliefs, desires, and other attitudes. Such judgments presuppose a variety of rational activities that together comprise a person's deliberations, such as the following: resolving contradictions among one's beliefs, working out the implications of one's beliefs and other attitudes, and ranking one's preferences in a transitive ordering. Each of these rational activities is directed at meeting a specific normative requirement of rationality, such as the requirements of consistency, closure, and transitivity of preferences. Deliberation involves many more rational activities, each of which is similarly directed at meeting some specific normative requirement of rationality. But it doesn't matter for my argument here what these activities happen to be. What matters is that all of these rational activities have a common purpose, which is to contribute to the overarching rational goal of arriving at and acting on all-things-considered judgments. If it is not evident to you that the more specific normative requirements of rationality do contribute to this overarching rational goal, try to imagine what it would be like to arrive at all-things-considered judgments without satisfying them. If you refused to resolve contradictions among your beliefs, for example, there might be no such thing as a best action for you to take in the light of your beliefs. It might be that one of your beliefs directs you to perform a certain action while its contrary (which you also believe) directs you not to perform it. Similar problems would arise if you refused to work out the relevant implications of your attitudes or to rank your preferences transitively. You would be refusing to consider all things in the sense required for deliberation; you would be refusing to consider their rational import. I'm going to call the state that would be achieved if a person were to succeed in this endeavor of arriving at and acting on all-things-considered judgments the state of *overall rational unity.*

Thus, persons, as rational agents who can engage in distinctively interpersonal relations, are subject to the normative requirement to achieve overall rational unity within themselves. Not only are they subject to this requirement but they must also be *committed* to satisfying it, because interpersonal forms of engagement always aim at rational response, and the only way to achieve this aim is by appealing to another person's own commitment to being rational. I've just unpacked this commitment as a commitment to achieving overall rational unity by arriving at and acting on all-things-considered judgments. And it should not be thought that this aspect of interpersonal engagement—that it appeals to a person's own commitment to being rational—is somehow compromised by the obvious and undeniable fact, discussed above, that persons are bound to fall short of being fully rational. My point there was that such failures are intelligible as failures only insofar as there are normative requirements of rationality to which we can appeal as standards of evaluation. The parallel point here is that it would make no sense to evaluate persons as falling short of those requirements unless they were themselves committed to meeting those requirements. Thus, the notion of commitment I'm introducing here is relatively weak. The following suffices for a person's being committed to meeting the normative requirements of rationality: (1) the person takes those requirements to articulate what it would be to be fully or ideally rational and (2) the person can recognize failures to meet the requirements as grounds for self-criticism and self-improvement.

As my argument proceeds to the next stage, it will not matter what the various specific normative requirements of rationality are, exactly. All that matters is that, taken together, these specific requirements serve a more general requirement on persons, which is to achieve overall rational unity within themselves. And the really crucial point for my argument is that *persons have a commitment to achieving such overall rational unity*, for this commitment is what enables persons to engage one another in distinctively interpersonal ways and thereby satisfy the ethical criterion of personhood.[5]

This elaboration of what it means to satisfy the ethical criterion of personhood makes an *implicit reference to personal identity*. The normative requirement to achieve overall rational unity *defines* what it is for an *individual* person to be fully or ideally rational. This can be seen from the fact that there is no failure of rationality when a group of persons fails to meet this ideal, only when an individual person fails to meet it. So, for example, if I have inconsistent beliefs, I am guilty of rational failure; but my beliefs may be inconsistent with yours without any rational failure on either of our parts, because you and I might simply disagree.

One promising way to approach the issue of personal identity, then, is to investigate the condition in which the normative requirement to achieve overall rational unity applies—or, rather, the condition in which a commitment to meeting the requirement arises. This is the condition in which we have a *person* in the sense that goes together with the ethical criterion of personhood.

It might seem that this way of approaching the issue of personal identity should bring us fairly swiftly to the conclusion that there is no distinction between personal identity and human identity after all. The condition of personal identity has been equated with the condition in which the commitment to achieving overall rational unity arises. In what condition does this commitment arise? When there is something that can grasp the normative force of the requirement to achieve such unity. In what condition can something grasp such normative force? When it possesses the requisite rational capacities. Which things have these rational capacities? The only known cases are human beings. Insofar as human beings are born with these capacities, it belongs to their nature to exercise them. And I have just argued that the only way to exercise them—the only way to be rational—is by aiming to achieve overall rational unity. So it might appear to follow from my argument that it is in the nature of a human being to be an individual person in exactly the sense that goes together with the ethical criterion.[6]

However, this does not follow. Although rational capacities must always be directed at achieving rational unity somewhere, they needn't be directed at achieving rational unity within the biological boundaries that mark one human being off from another. Human beings can exercise their native rational capacities so as to achieve different levels of rational unity within different boundaries. They can exercise their rational capacities together so as to achieve rational unity within groups that are larger than a single human being, and they can exercise their rational capacities in more restricted ways so as to achieve rational unity within parts that are smaller than a single human being. When either of these happens, it is not individual human beings but, rather, groups or parts of human beings that can engage in distinctively interpersonal relations and, hence, be treated specifically *as persons*.

These claims are bound to meet with some skepticism. Unfortunately, I can't give a full defense of them here. What I can do is indicate the kinds of consideration that support them and, in doing so, further elaborate their meaning.

I'll start with the case of group persons. It is well known that when human beings engage in group activities, their joint efforts can take on the characteristics of individual rationality. Think, for example, of marital partners who deliberate together about how to manage their homes and families and other joint concerns. They may, in the course of such joint deliberations, do as a pair all of the things that individuals characteristically do to be rational: they may pool their information, resolve conflicts, rank preferences together, and even arrive at all-things-considered judgments together about what they should together think and do—where the "all" in question comprises all of their pooled deliberative considerations. The same can also happen in a less thoroughgoing way when colleagues coauthor papers, or when teams of scientists design and run experiments together, or when corporations set up and follow corporate plans. We tend to assume that such joint endeavors leave human beings intact as individual persons in their own rights. Insofar as that is so, it should be possible to engage those human beings separately in conversation, argument, and other distinctively interpersonal relations. But, sometimes, this is not possible. Sometimes, marital partners won't speak for themselves. Their commitment to deliberating together is so thoroughgoing and so effective that everything they say and do reflects their joint deliberations and never reflects their separate points of view. The same can happen to coauthors, team members, and bureaucrats. The kind of case I have in mind is not one in which human participants simply wish to give voice to the larger viewpoint of the group to which they belong, but one in which the human constituents of the group are not committed to having a separate viewpoint. That is, these human beings are not committed to achieving overall rational unity separately within their individual lives. Yet this is not because they lack rational capacities. It is because those rational capacities are directed in a different way, so as to generate a larger commitment on the part of a whole group to achieve overall rational unity within it.

If this seems implausible, just think about two different attitudes you might bring to a department meeting. You might bring your own separate viewpoint to the table, with the aim of convincing your colleagues to agree with you. This attitude takes for granted the status of each colleague as an individual person in her own right with her own separate point of view. The attitude also *perpetuates* that status, for the effect of adopting it will be that you maintain the separateness of your point of view by deliberating on your own, with the aim of achieving rational unity just within your own self. Even when you are moved by what your colleagues say, the reason is not that you want to resolve disagreements with them or do anything else that would help you achieve rational unity as a group. You will be moved by your colleagues only insofar as what they say bears on your personal project of achieving such unity by yourself—by showing you that you have internal reasons, from your own point of view, to accept what they are saying. But you might bring a quite different attitude to a department meeting, one that would not perpetuate your separateness from your colleagues. You might bring to the table all you have thought of with respect to the issues the department faces, with a view to pooling your thoughts with your colleagues' thoughts, so that you can together discover the all-things-considered significance of the whole group's thinking. If your colleagues do the same, then it won't be true that each of you is committed to achieving overall rational unity on your own; there will be a joint commitment on the part of the whole department to achieving such unity within itself. And, for this reason, it will be possible for others to engage the department in conversation, argument, and other distinctively interpersonal relations. The department could be asked, for example, why did you do such and such? And there will be a coherent answer that reflects the department's joint deliberations. I'm not saying that departments of philosophy typically have the commitment that would render them sufficiently unified to be engaged in this way. But I am saying it is a possibility. It is possible for human-size philosophy professors to undertake a commitment to achieve rational unity together. And, if they did undertake such a commitment, then the lines that divide one person from another would have shifted. They would no longer follow the biological divisions that mark off different human beings. Nor would they follow the phenomenological divisions that mark off one center of consciousness from another. They would follow nothing else than the commitment to rational unity that is characteristic of the individual person. *To say that these lines can be redrawn in these ways is to say that the facts of personal identity are matters of choice, not a metaphysical or biological given.*

Many objections could be raised at this point, more than I have space to address here. I'll briefly take up two before moving on to the case of multiple persons. These objections do not call into question whether human beings can coordinate their deliberative and practical efforts in the ways I've been suggesting. What they call into question is whether such group endeavors could ever undermine the status of individual human beings as persons in their own rights.

The first objection notes that my descriptions of group persons referred at certain crucial points to the thoughts and choices of their individual human participants. For example, I referred to two different attitudes that an individual philosophy professor might bring to a department meeting. One would maintain its internal rational unity and, thereby, the separateness of its point of view, while the other would contribute to the overall unity of the department and, thereby, help constitute the department's more inclusive point of view. I also described two marital partners as each being committed to engaging in joint deliberations. My language may have given the impression that the unity of the group person is, in each case, actively maintained through individual commitments and efforts on the part of its human members, the philosophy professors and marital partners, respectively. And this seems to imply that the human members themselves must remain individual persons in their own rights even in the context of a group endeavor, because otherwise they could not possibly maintain group unity through their individual commitments and efforts. But my language was misleading. What is true is that a group person may *initially* be brought into existence through the individual decisions and actions of smaller persons, typically of human size. But if these initial efforts are successful, then a group person will be brought into existence. And, thereafter, at least some of the intentional episodes that occur within the human organisms involved will be episodes in the life of a group person rather than in the separate lives of human-size persons. Going back to the case of a philosophy department: when I bring to the table the aim of joining in a departmental deliberation, then, insofar as my aim is shared by others and is efficacious, the result will be that the subsequent deliberations *around* the table are carried out from a new, emergent group point of view that can't be equated with my point of view or any other human-size point of view. This is not to say that there is no sense in which separate, human-size points of view would be left intact. All of the human beings involved would still have separate centers of consciousness and, hence, separate phenomenological points of view. But it is important not to confuse the idea of a *phenomenological* point of view with the idea of a *rational* point of view. The latter is the point of view from which an agent deliberates, and it is a point of view that others can engage in distinctively interpersonal ways by virtue of its commitment to achieving overall rational unity within itself. In the case under discussion, the separate phenomenological points of view of the human members of a unified philosophy department do not qualify as separate rational points of view in this sense. Such centers of consciousness are not centers of rational activity aimed at achieving rational unity within them. They are mere sites of rational activity aimed at achieving rational unity within the larger boundaries of the department. When this occurs, the human members of a group can no longer be engaged as individual persons in their own rights. And, so, the fact that they remain intact as individual animals with separate phenomenological points of view does not suffice to show that they also remain individual persons, as the objection alleges. (This, of course, is why I was so insistent in my introductory remarks that we take care *not* to gloss over—as Locke did—the distinction between *consciousness* and *rationality*. This is so because of the way in which centers of consciousness need not fall one-to-one with respect to centers of rational deliberation.)

The second objection pursues a line of thought similar to the first, but somewhat more aggressively. This objection insists that it is always possible to describe the phenomena associated with group endeavors in methodological individualist terms—that is, in terms of thoughts and actions that belong to the human participants. The objection concludes that there is no reason to grant my claim that group endeavors may literally alter the boundaries between persons. My response to this objection is that it may be carried even farther, farther than its proponents intend. If it is possible to describe group endeavors in methodological individualist terms, it is equally possible to describe human endeavors in homuncular terms. So, if it is supposed to follow in the one case that human activity cannot produce group persons, it should follow in the other case that homuncular activity cannot produce persons of human size. But certainly, the latter inference does not go through. Homuncular activity can be directed in such a way as to yield overall rational unity within the larger human being. Indeed, this had better be so. Otherwise, the homuncular theory would not be consistent with known facts about human social interactions, which often take the form of interpersonal relations in which one human-size person actively engages the rational point of view of another human-size person. This would not be possible unless homuncular activity could be directed at achieving overall rational unity within the whole human being. And, if this can happen among homunculi, it can also happen among human beings: their rational activities can be directed at achieving overall rational unity within a whole group.

It may seem unfair to have saddled the proponents of methodological individualism with homuncular theory. But this is not unfair insofar as methodological individualism is supposed to follow from the fact that it is always *possible* to describe group endeavors in methodological individualist terms, and thus it is relevant that it is *also* possible to describe human endeavors in homuncular terms. Now, it is often observed that references to homunculi are *not necessary* in adequate descriptions of human psychology. And that may be the spirit in which methodological individualists rule out the existence of group persons; they may be claiming that references to group persons are not necessary in adequate descriptions of human psychology, any more than homunculi are. So, by Occam's razor, the only persons left standing are of human size—nothing larger or smaller need be posited. But this reasoning begs the question in favor of the basicness of human beings. From the point of view of homuncular theory, references to human-size persons might seem just as unnecessary as references to homunculi might seem from the point of view of a humanistic theory. We can easily avoid this impasse by returning to the ethical criterion of personhood, which supplies a useful pragmatic test. By the lights of that criterion, something counts as a person if it can be *effectively treated* as a person through distinctively interpersonal forms of engagement. Homunculi don't pass this test. Many human beings do. And the latter should not be disqualified from the status of persons just because clever theorists can describe their mental lives in homuncular terms. Similar remarks apply to group persons. They qualify as persons insofar as they pass the test of being treatable as persons. And they should not be disqualified from the status of persons just because clever theorists can describe their intentional activities in methodological individualist terms. All that matters in either case is whether there is a requisite commitment to rational unity, by virtue of which it is possible to engage them—human beings or group persons—in distinctively interpersonal ways.

Multiple persons are distinguished from homunculi precisely by the fact that they satisfy the pragmatic test of personhood that follows on the ethical criterion; they can be treated as individual persons in their own rights, even though they cohabit the same human body. Let me now outline the considerations that I think support the idea that such multiple persons are possible.

The considerations are really generalizations from the case of group persons. That is, I propose to model *all* cases of rational unity on the unity of a group. My suggestion is that rational unity doesn't *just happen* as the inevitable product of some natural process, such as the natural biological development of a human being. Rational unity is something that is *deliberately sought* for the sake of some *further end*. There are things that a philosophy department can do as a unified group person that no human-size person can do on its own. And that may constitute a *reason* for which such human-size persons might initially decide to pool their efforts in a joint endeavor. If they implement their decision, they may no longer maintain separate rational points of view. So, what perpetuates the group person once it has been brought into existence is not human-size persons' separate commitments to it; it is up to the group itself to maintain its existence by continuing to strive for overall rational unity within it.

When we view the unity of a human-size person along these lines, we must see it as deliberately achieved for the sake of some further end that couldn't be achieved without it. The appropriate contrast here is with an impulsive human being who doesn't strive for rational unity—who doesn't deliberate but simply follows current desires unreflectively and uncritically. Because the capacity to deliberate belongs to human nature, perhaps it is fair to say that such a human being is acting against its nature. But that doesn't harm my point, which is that when human beings do exercise their rational capacities, they are *striving to achieve* rational unity through their intentional efforts. And it is part of this same point that these capacities can be directed at the achievement of rational unity within different boundaries. An initially impulsive human being might, in time, strive for rational unity within each day, or week, or month, or year, or even a whole lifetime. The last goal was celebrated by Plato as part of the just life and by Aristotle as part of the virtuous life. In a less high-minded way, we now typically pursue the project of living a unified human life for the sake of other, more specific projects, such as life-long personal relationships (friendships, marriages, families) and careers. But what I want to emphasize is that these are *projects* and they are *optional*. It is possible for human beings to strive for much less rational unity than these projects require and still be striving for rational unity. And, sometimes, the result may be relatively independent spheres of rational unity with a significant degree of segregation.

Such segregation is evident to some degree in the lives of many human beings whom we find it possible to treat, for the most part, as roughly human-size persons. We may find, for example, that when we visit corporate headquarters our friend "becomes" a bureaucrat who cannot recognize the demands of friendship. What this means is that our friend's life takes up a bit less than the whole human being we are faced with, the rest of which literally belongs to the life of the corporation. According to the account of personal identity that I'm now elaborating, this may not be mere "role playing." This may be, literally, a fragmentation of the human being into relatively independent spheres of rational activity, with separate rational points of view that can be separately engaged. Of course, group endeavors do not necessarily result in such fragmentation; they might completely absorb the human lives that they involve (this may happen in the armed forces and in certain intense marriages). But when a group endeavor does *not* completely absorb the human lives that it involves, there is a consequent split in those lives. And I propose to conceive multiple persons along precisely these lines. The only difference is that the separate rational points of view of multiple persons need not be imposed by involvements in group projects but, rather, by involvements in other sorts of projects that a single human being cannot pursue in a whole-hearted and unified way. When a human being's projects are numerous, and when they have nothing to do with one another, this may make it pointless to strive to achieve overall rational unity within that human life. And the rational response may be to let go of the commitment to achieving such overall rational unity within that human life and to strive instead for as many pockets of rational unity as are required for the pursuit of those relatively independent projects. So, just as a group person may dissolve itself for the sake of human-size projects that would otherwise have to be forsaken for the sake of the group's overall unity, a human-size person may dissolve itself for the sake of even smaller projects that would otherwise have to be forsaken for the sake of the human being's overall unity. In such conditions, we find the emergence of multiple persons within that human being, each of which can be treated as a person in its own right.

I don't want to suggest that this is typically how persons come into existence, through the breakdown of some larger unity. I think it usually occurs in the reverse direction: persons notice that there is something worth doing for the sake of which more unity needs to be achieved. That is certainly how group persons would typically come into being. And I'm suggesting that the same holds for human-size persons.

In conclusion, then, the reason for my making a distinction between personal identity and human identity is related to the role of rational agency in personal life. Even though human beings are, in the normal case, born with the sorts of rational capacity that make personhood possible, the actual exercise of their agency is required for human beings to become individual persons. And human beings have it within their power to direct their agency for the creation of persons who are larger or smaller than human size.

The Four Cases

I'll treat cases 1 and 2 together. My account implies that human beings such as Peter Jones and Daniel Smith do not qualify as persons during advanced stages of Alzheimer disease (AD) or other dementia. Although they may be capable of limited forms of conversation, they are not capable of entering into the full panoply of interpersonal forms of engagement. They cannot be held responsible for their actions. More poignantly, they also cannot maintain personal relationships that require responsiveness to reasons.

My account also brings out that there is no pressing metaphysical reason to say that, in these cases, the life of the person continues into the times when the capacity for rational agency and the related capacity for interpersonal engagement have been lost. It makes far more sense to say that when the exercise of these capacities comes to an end, so too does the life of the person. The matter would, perhaps, look different if the possibilities of group and multiple personhood were not in view. If we were to bracket these possibilities, then it would not be so clear that the existence of a person depends on the exercise of rational capacities. It might appear instead that the existence of a person depends only on the existence of an organism whose nature is to have such capacities. On this view of the person, persons are human beings, and there is no distinction between personal and human identity. Rather, personhood is a status that a human being might gain or lose at different points in life, in the way I described at the start of the last section, depending on whether it possesses and is able to exercise rational capacities. But the existence of the person, as human being, would be a *biological fact* and, in that sense, a *metaphysical given*. However, once we recognize the possibilities of group and multiple personhood, it doesn't make sense to view the person in this way. We have to view it as something that emerges through the exercise of human rational capacities. It is precisely through the exercise of such capacities that individual persons can emerge within different boundaries—boundaries that are sometimes larger than, sometimes smaller than, and sometimes roughly the same as the boundaries of the individual human life. When we think of persons as *emergent* in this way, it seems right to say that AD or other dementia marks the death of a person even though it does not mark the end of a human life.

If there were any reasons to deny this, they would have to derive from ethical considerations rather than from strictly metaphysical considerations. So let me briefly consider why it might seem ethically important to deny this.

Let me remind the reader that we do not need to view individuals with AD or other dementia as persons to justify ongoing treatments aimed at improving health and prolonging life. It would be enough to note that they are human beings and to give an account of why we believe in the sanctity of human life. What I am considering here is something different. I'm considering whether there are reasons that it might seem somehow inadequate to register that these patients are *human beings*—reasons that it might seem necessary and important, for certain ethical purposes, to say that they are *persons*.

Much of our current thinking about rights and obligations is organized around the anti-Lockean idea that a person's life continues until biological death and does not end with the loss of rational capacities. For example, a person who anticipates AD or other dementia might claim a right to make advance care directives. It is natural to conceive this right as a right to self-determination.

And, prima facie, we cannot think of the right to *self*-determination as extending to times after competence has been lost unless we think of *the life of the self* (i.e., the person) as extending to those times as well. But this is not something that my version of Locke's distinction allows us to say. Only an anti-Lockean, who equates the life of the person with the life of the human being, can portray advance care directives made in anticipation of AD or other dementia along these lines, as acts of self-determination. (This anti-Lockean conception of personal identity also seems to supply a plausible account of our *motives* for advance care directives: the main reason that we want to make such directives is that we believe it is *our own future* that is at stake.) Another area in which it might seem ethically important to conceive personal identity as coinciding with human identity concerns promises and contracts. We might want to insist, for example, that our marriage not be dissolved and that our property not be disposed of until biological death, even if we should become incapacitated by AD or other dementia before that time.

Although much current thinking about rights and obligations certainly is organized around the anti-Lockean conception of personal identity, in the ways I just described, I now want to bring out that it need not be so organized. In fact, it would be better if it were not so organized.

With respect to advance care directives, we should bear in mind that we already recognize the right to make care directives for others besides our selves —namely, other human beings who fall within the domain of our responsibility, such as our children and other dependents. Because this is so, our right to make advance care directives does not require us to suppose that it will be our *very selves* who will suffer when our personhood is no longer in place. We can conceive such directives as arranging for the future well-being of human beings who fall within our care, not because they are *our selves,* but because they were once the sites of our personal lives.

The case of marriage is more complicated. Obviously, a marriage cannot outlive its partners. It follows that if a marriage can continue after one or both partners have lost certain rational capacities, then, likewise, it must be the case that its partners can continue to exist after such losses. However, it does not follow that a *person* can continue to exist after such a loss. What follows is that marriage is not always a relation between persons. It can also be a relation between human beings. This is certainly part of the traditional conception of marriage, conceived as a relation between a man and a woman who share the project of having and raising children together. (Thus, in the context of a traditional marriage, one's spouse remains one's spouse regardless of whether it is a person any more, much as one's child is one's child regardless of whether it is ever a person.) Of course, this is not the only way to conceive of marriage and marital partners. Marriage can also be conceived of as a legal contract between persons. So conceived, a marriage would effectively be ended as soon as either partner ceased to be legally competent. But it is important to bear in mind that the issue here is not merely legal. For many of us, the point of marriage is to share in activities and decisions in a way that is available only to persons. If that is the point of a marriage, then, apart from legal issues, it probably ought not to be conceived of as remaining in effect once personhood is lost. The upshot of these reflections on marriage is as follows. On the one hand, we can think of marriages as continuing when partners have advanced stages of AD or other dementia as long as we are prepared to think of marriage as a relation that holds between human beings. But, on the other hand, this leaves scope for embracing one or the other version of Locke's distinction and allowing that a person may die a nonbiological death through psychological loss. It also leaves scope for conceiving of marriage as a relation between persons in the strict sense that I have been defending in this chapter. In this conception, a marriage might be ended by the nonbiological death of one (or both) of its partners, through the loss of rational capacities.

I'll close my discussion of cases 1 and 2 by returning to a point that I made in the first section, concerning rights and obligations. Just as there is a distinction between those things that can be treated as persons and those that cannot, there is a corollary distinction concerning those rights and obligations that attach to persons and those rights and obligations that attach to nonpersons. The former can be lumped together under a more general right to self-determination, and these rights and obligations include the basic freedoms associated with the liberal political tradition, such as freedom of speech and conscience. In addition to these freedoms, some personal rights that have figured centrally in liberal political theory include the right to enter into contracts and the right to vote. Examples of other rights that need not be conceived as *personal* rights would include rights to quality of life, to medical care, and to other kinds of care as well. It is significant that these latter rights can coherently be extended to other animals besides human beings, animals that clearly lack the capacity for personhood altogether. And this should help make my general point here vivid. We should not make the mistake of supposing that these latter rights and obligations require us to view nonpersons as persons. Nor should we make the mistake of supposing that they require us to view the life of the person as extending through an entire human life span, to times when the capacity for interpersonal engagement has been lost. Those who have AD or other dementia may have rights, and we may have obligations to them as long as they live. But these are rights we award to them as human beings, not as persons.

A great deal more needs to be said here. For one thing, it is not immediately clear what my general point implies with respect to the issue I raised above concerning whether and why marital obligations and property rights can continue to hold in advanced stages of AD or other dementia. But a certain amount of confusion is to be expected. Institutions such as marriage and property are designed in the first instance by and for persons, and they make only partial sense in application to nonpersons (such as human beings in advanced stages of AD or other dementia). Moreover, once we rethink the issue of personal identity along the lines I have been proposing, it is inevitable that we have to rethink the whole matter of rights and obligations. Not only will we have to be clearer about the difference between the sorts of rights and obligations that can attach to persons and to nonpersons, respectively, but we will also have to be clearer about what sorts of rights and obligations, if any, can attach to multiple and group persons. (As an aside, I should mention that recognizing the possibility of group persons might constitute a salutory first step toward an adequate account of corporate responsibility.)

I turn now to cases 3 and 4, in which deep brain stimulation (DBS) and steroids induce drastic personality changes. As should be evident from my account, personhood remains very much in place throughout the human lives described in both of these cases. But the account does put into doubt whether one and the same person continues to exist throughout each of those human lives. I have argued that the existence of an individual person arises with a commitment to achieving overall rational unity within specific boundaries that mark that one person off from others as something with a separate rational point of view. And I have argued, further, that these boundaries are set by substantive commitments to projects that give persons reason to achieve unity with them (rather than, instead, within larger or smaller boundaries). Persons of roughly human size arise, then, only insofar as they have projects that require them to achieve overall rational unity within a particular human life. And if such roughly human-size persons are to persist over time, their projects must require rational unity over time as well as at a time. In life as we presently live it, these projects typically take the form of relationships and careers. Our relationships and careers provide us with reasons to achieve significant unity over time within a single human life span. And so it is in the lives of the persons described in cases 3 and 4 before they begin the treatments that lead to their personality changes. The question to which my account of personal identity directs our attention concerns whether the treatments eventually undermine the commitments by virtue of which these persons continued to exist. Because there are important differences of detail in the two cases, I'll discuss them separately. As I discuss them, I'll bring out some further aspects of personal identity and personal agency that I haven't yet had occasion to mention.

My account provides little basis for supposing that, in case 3, the person who emerges from DBS is identical to the person who had, before the treatment, enjoyed his career and marriage. After the treatment has its effect, the person called "Charles Garrison" more or less sets aside those old projects in favor of new ones. It is perhaps less clear that he abandons his marriage than that he abandons his former career. But as the case is described, his marriage seems to figure more as a background fact than as a commitment. This is signaled by the way in which he doesn't bother to take into account the possible consequences that quitting his job might have for his marriage. One important general criterion by which to judge the discontinuity of commitment here concerns his willingness to identify with and take responsibility for earlier actions and values. By the end, he has undergone such a profound alteration in his general outlook on the world that it seems unlikely that he could identify with his past actions and values. For this reason, it is unclear that it would be meaningful to try to hold him responsible for them. All of these considerations incline me to say that he is, in the end, a distinct person from the person described at the beginning of the case.

However, a delicate issue arises here and will arise again in connection with case 4. It concerns the basis on which we verify the presence of commitments. (To clarify: when I speak of *commitments* I am referring to both the general commitment to overall rational unity that characterizes the life of any person and, also, specific commitments to particular projects that set the parameters of that general commitment in individual lives.) A commitment is not, straightforwardly, a disposition to behave in accord with it. A commitment also has an irreducibly normative dimension. A person's commitments tell it what it *ought*, by its own lights, to do. An ideally rational agent would live up to its commitments—that is, it would do exactly what it ought (by its own lights) to do. But, as I brought out in my discussion of rational agency in the second section, no real persons (who are not God or angels) are able to keep themselves perfectly in line with their commitments. When we fail to live up to our commitments, this does not necessarily mean we have lost them. We may still retain them. And when we do retain them, they provide a normative standard in the light of which we are able to think about what we ought to have done but didn't do. It is this normative role that distinguishes commitments from mere dispositions. The delicate issue, then, concerns how to discern whether a person has certain commitments when the person's actions and behavior are not in accord with them. The surest evidence is this: that the person is prepared to accept criticism for not being in accord with them. Preparedness to accept criticism for not living up to a commitment reveals that the commitment is still playing a normative role in the person's life—even in cases in which the person's causal dispositions are not otherwise in accord with the normative requirements of the commitment.

The case of Mr. Garrison brings out clearly just how delicate an issue it can be to attribute commitments in this normative sense. When he begins to develop apathy, associated with Parkinson disease, his behavior is clearly not in line with his previous commitments to his job and family. But the question is, does it make sense to suppose that he still has those commitments nonetheless and that the disease is simply preventing him from living up to them? Here is what stands in the way of an affirmative answer. Not only does he fail to behave in accord with his earlier commitments to his job and family, but there is no evidence that these commitments are functioning in his life as a normative standard, insofar as he shows no guilt or other self-critical response when he fails to act in accord with them. However, there is one remaining —albeit slim—basis on which we might suppose that he still has his earlier commitments: if we thought that the purpose for which he decides to undergo DBS to begin with is to enhance his chances of living up to those commitments. Unfortunately, the description of the case does not provide us with clear evidence that this is indeed the purpose for which Mr. Garrison undergoes DBS. The decision to go ahead is described as a collective one made by the entire family, and we are not told what his own attitude is. It seems likely that his family members would view the treatment as a way of trying to save his original self from destruction by Parkinson disease. Although I can't say for sure that their view would be wrong, I have my doubts. In the interim period when apathy sets in, it isn't really clear that there remains an original self still to be saved—precisely, there is no definite sign of any ongoing commitments that span the past as well as the present.

A similar lack of clarity arises in case 4, as told with the first ending. But let me first quickly respond to the case as told with the second ending, which is more straightforward. At the outset, John Fast does not intend to induce a personality change, nor does he intend to lose his commitments to his career and family. On the contrary. His initial reason for taking steroids is to promote his career, and it seems he does not really believe that taking them would undermine either his career or his home life. This strongly indicates that all of his central and unifying personal commitments are in place when he begins taking steroids. With ending 2, the same commitments clearly are still in place when he decides to stop taking steroids, because he stops taking the drugs precisely to live up to those commitments. It makes sense to suppose that the commitments remain in place in the intervening period as well, even though he isn't living up to them. This provides the best account of why he eventually agrees to stop taking the steroids. So, by my lights, he qualifies as one and the same person throughout the entire period described in the case as it is told with ending 2.

However, this conclusion is not so clearly implied by ending 1, in which he refuses to stop taking steroids even when he is threatened with the loss of both his sports contract and his marriage. In this telling of the case, there is room for multiple interpretations. The most obvious interpretation says that, at some point after beginning steroids, he loses virtually all of his earlier commitments. Although he still seems to care in a general way about being an athlete, he seems not to care that his use of steroids threatens to undermine his career and marriage. This constitutes fairly strong evidence that he no longer has the commitments by virtue of which he would be identical, according to my account, with the person described at the beginning of the case, before the steroids have had their effect.

It might be argued that we can set this evidence aside and, moreover, that I must set this evidence aside if I am to arrive at a consistent view of both versions of the case, for there is a stage at which both versions seem to posit the same facts and yet I seem to be taking a different view of them. I am referring to the stage before Mr. Fast's official diagnosis, when he is first referred to his team psychiatrist. At that stage, he is portrayed as seeming to care very little about the negative consequences that his use of steroids is having on his career and marriage. When discussing the second version of the case, I was prepared to say that he nevertheless does retain his commitments to his career and marriage at that stage. That is, I was prepared to say that his apparent lack of interest in living up to those commitments is *not* strong evidence that he has lost them. Wouldn't consistency require me to say the same thing about him when, with ending 1, he continues to show the same lack of interest? There is a temptation to say yes here. To say yes is to take the view that there is an underlying self that abides throughout the case, which is somehow masked or distorted by the use of steroids. This view is reinforced by our knowledge that if he stopped taking steroids he would revert to behavior that accords with his earlier commitments—which, in my view, constitute his identity. Thus, we could say that one and the same person persists in both endings of the case, because the requisite commitments do remain in place; and we could say the difference between ending 1 and ending 2 lies only in how well the person lives up to those commitments.

But this is not what I think we should say. I think it is unwise to attribute commitments unless we find significant evidence that they are there. Although it would be too much to require, by way of evidence, that a person's behavior actually accords with his commitments, it would not be too much to require that a person show some sort of self-critical response when he fails to live up to those commitments. And such a self-critical response is nowhere in evidence in the case with ending 1. In contrast, it is very much in evidence with ending 2.

I'd like to close my discussion of the cases by situating my conclusions about cases 3 and 4 in relation to my larger claim that personal identity is always an achievement that requires effort and will—that one's identity doesn't just befall one as it would if it were a metaphysical given, a fact of biological (or other) nature. It may seem that this claim is not really consistent with my interpretations of case 3 and of case 4 with ending 1. In those cases, certain personality changes take place without being anticipated, let alone willed. And I have been suggesting that those changes may obliterate old commitments and usher in new commitments in their place and that this may mark the end of one person's life and the beginning of a new one—a nonbiological death and birth, as it were. It would seem that, in these two cases, the births in question are precisely *not* the product of effort and will. The emergence of these new persons looks like something that merely *happens* to them due to causal factors outside their intentional control—the causal effects of DBS and steroids, respectively. And this appears to contradict my claim that personal identity is always an achievement involving effort and will. However, there is no real contradiction here. The appearance of there being a contradiction derives from a false picture of personal agency and its role in personal identity. But this will take some explaining.

I'll begin with a point about agency that I take to be relatively uncontroversial. To say that something is a product of personal agency is to say that it is an achievement that involves effort and will. But this is not to say that it is due solely to the person's intentional activity, without any causal contribution from outside the operation of the person's own will. The idea of such a pure form of agency that operates in complete independence from all other causal influences is a fiction, at least for those of us who don't believe in an all-powerful being, such as the Judeo-Christian God portrayed in medieval theology. But even if we are prepared to elevate the idea of such a pure and independent form of agency from fiction to theological mystery, the more important point is that the idea cannot have any application in our own lives. And this makes for an interesting point of contrast with the requirements of rationality. I have pointed out that only God or angels, not finite agents like us, could fully satisfy these requirements. But all the same, these requirements do have application in our lives. It is meaningful for us to embrace them as ideals that we should strive to meet even though we never can, because even as we fall short of meeting them, they supply us with normative standards by which we can critically appraise our efforts to do so. There is no parallel role that the idea of a completely pure and independent form of agency can play in our lives. Not only is it the case that we couldn't ever exercise our agency in a way that was completely independent of all external causal influences, but there is no point in striving to exercise our agency with as little such causal influence as possible. That would simply deprive us of all the positive influences that the world, especially other persons, might have on our efforts to exercise our agency. And this serves to show that the idea of a completely pure and independent form of agency does not supply us with a normative ideal that we can coherently strive to meet.

Here, then, is the correct picture of the role of personal agency in personal identity. When I say that personal identity is a product of effort and will, I am not suggesting that persons make themselves from nothing, as if they could exercise their agency without any influence from the outside world, or even strive to do so. What I'm suggesting, rather, is this: whenever there is something whose substantive commitments give it reason to achieve rational unity within certain boundaries, there is an individual person, in my sense; and, because persons are always open to the possibility of redrawing those boundaries, their continued existence as the persons they are is always something they are actively choosing. *This is why personal identity is an achievement involving effort and will.* It is not because persons have already, in the past, chosen everything that is presently true of them. It is because their identities must always be actively maintained in favor of available alternatives.

It may not be evident from what I have said so far how the correct picture of personal agency helps sort out the proper interpretation of cases 3 and 4, and, especially, the significance of the two different endings of case 4. Indeed, there is much more that needs saying. But I hope this much is evident: we need not, and should not, think of persons' commitments as things wholly of their own making, without any causal contribution from the outside world. And this opens the door to the following thought: when patients undergo various medical treatments, and when they seem to gain new commitments in place of old ones as a result of those treatments, we shouldn't dismiss this as *mere* appearance, simply on the ground that the patient didn't directly or wholly control the process by which it happened. Agents never do have such total control over what we think of as the products of their own agency.

Some might protest that this thought overlooks what is really the crucial point about agency. While it is surely true that agents like us exercise our agency *in tandem* with outside causal forces, rather than independent of them, there must nevertheless be a *distinctive* causal contribution that we make when we exercise our agency, for, otherwise, we will lose our grip on the very distinction between the products of our agency, on the one hand, and mere happenings, on the other. We ordinarily conceive such distinctive causal contributions of agents in terms of the *reasons* on which they act. And here it is useful to construe *action* in the broadest possible way, to include what might be thought of as mental actions, such as the adoption of commitments, as well as bodily actions—for in both cases, it seems that we perform actions for reasons. It is, of course, precisely such reasons that seem to be missing in case 3 and case 4 (with either ending). Mr. Garrison and Mr. Fast seem to come by their new commitments without undertaking them for reasons, but simply as causal by-products of DBS and steroids, respectively.

Let me explain, then, why I think we ought not to place too much emphasis on the causal etiology of actions and commitments—and, in explaining this, further elaborate what I see as the correct picture of personal agency. First of all, much of what we hold persons accountable for, in Locke's sense, is not properly described as an action undertaken for reasons. A great deal of our intentional behavior simply doesn't fit that description. Much of it is habitual or automatic in some other way—or, if not automatic, still not the product of explicit reasons arrived at in the course of deliberation. We are still prepared to count our behavior as intentional and as something for which we are accountable, as long as we have relevant commitments in the light of which we find it appropriate to evaluate our behavior. So, for example, my failure to exercise today can be seen as something that was deliberate and intentional, not because I sat down and decided not to exercise for reasons, but because I have a general commitment to exercising every day in the light of which I can criticize myself—and in that sense hold myself accountable—for not exercising today. So what makes the difference between actions and mere happenings profound is not exactly causal etiology, but that actions bear a normative, rather than a causal, relation to an agent's underlying commitments.

This point has important implications for our attitude toward various kinds of intervention, treatment, and therapy. I have insisted that it is possible to have commitments even if one doesn't have the ability to live up to them, as long as one is willing to criticize oneself for one's failures to live up to them and to take whatever measures might be available for doing better. All sorts of treatment, including DBS, steroids, and other drugs, are to be counted among such measures. Viewed in this way, these treatments are not a substitution for the exercise of personal agency but, rather, ways of exercising it. In some cases, they may be the best—and, indeed, the only—way to live up to one's commitments.

It must be admitted, however, that this point is, in the first instance, a point about *bodily* action. The behavior that a person exhibits need not be caused by choice and intention to come under the normative scope of evaluation in the light of the person's commitments. Drugs and treatments (as opposed to choices and intentions) may cause a person's behavior and, still, we can ask whether the behavior is or is not in accord with the person's commitments. But could similar reasoning possibly apply to commitments themselves? Perhaps it does matter how commitments come about. Perhaps the important thing about commitments *is* their causal etiology—that they are undertaken for reasons and are not merely the causal effects of certain treatments and other causal processes that we undergo.

I agree that, in the ideal case, we would arrive at our commitments for reasons, in a sense that implies they are the direct causal products of our reflections and deliberations. But we are all well aware that this is not the only way our commitments come about. And here I'm not referring to how certain medical treatments can cause a person's commitments to change. I am referring to the operations of character and taste and other obscure causes that are always at work in the lives of persons. Speaking for myself, some of my most cherished commitments—including my commitment to philosophy—are due to hidden causes that I shall probably never discover.

What makes a commitment a commitment is not its causal etiology but rather the normative function that it fulfills in our lives—that it functions as a standard by which we evaluate our behavior as being in or out of accord with what we think we ought to do. And of course, there is a parallel function within the circle of our commitments themselves. Just as commitments serve as the standards by which we evaluate our behavior, likewise, they serve as the standards by which we evaluate *them*. This may seem circular. And so it is. But we should not aspire to get outside the circle of our commitments, to stand above or apart from them so that we can evaluate them all at once. This would be a psychological impossibility. And more directly to the point here, it would also be a normative impossibility. Our actual commitments are the only normative basis we have for evaluating our commitments. The most we can do is evaluate them piecemeal, by taking some for granted and using them as the measure of the rest. And there is no reason that the commitments that arise through medical treatments should be deprived of this normative role. In fact, I don't see how we could possibly prevent patients from according them this normative role, except by *further* medical interventions—and then the same problem would arise. At some point, we have to allow that the actual commitments a patient has are bona fide commitments, no matter what their causal etiology might happen to be.

How, then, should we view the cases in which we can *foresee* that a given medical treatment is likely to cause a person's commitments to change? Clearly, when patients and doctors deliberate about whether to pursue such treatments, this is a fact that must be weighed, alongside all of the other facts about the costs and benefits likely to follow as a result of the treatment. Someone who anticipates the progression of Parkinson disease might well prefer to undergo profound alterations of commitment as a result of DBS rather than let the disease bring on apathy. Similarly, an athlete might prefer (albeit less reasonably) to undergo profound alterations of commitment as a result of steroids rather than accept his or her current physical limitations. If these preferences are intelligible, this means that persons can have *medical* commitments (e.g., to preventing the effects of Parkinson disease or to overcoming physical limitations) in the light of which they are willing to allow that many other new commitments will befall them. And they can coherently do this even if they can foresee that undergoing medical treatment will involve losing all of the other commitments that currently shape their lives and constitute their identities. *On my account of personal identity, if patients choose to undergo medical treatment in the expectation that profound changes in their other commitments will follow, they will be choosing to bring about their own nonbiological death, to be followed by the nonbiological birth of another person whom no one yet knows.*

Summary

Let me close by bringing the several parts of this chapter into more perspicuous relation. I have just argued that we ought to view certain medical decisions as raising the possibilities of nonbiological death and birth. To view them in this way is to take up the perspective from which the possibilities for which I argued in the second section—of group and multiple persons—are also visible. I argued there that it is within the power of human-size persons to decide that there are things worth doing for the sake of which it would be better to integrate into larger persons or to fragment into smaller persons than to continue as who we are. And this is how I am recommending we view medical decisions to undergo treatments that will cause significant psychological change. In some cases, the decision to undergo treatment will be a decision to fragment. It will be a decision to cede to someone else's existence within our own body rather than to continue as who we are. This need not imply that we ought not to undergo such treatment. It all depends on whether we think that something is worth doing that, in our estimation, could be done by the person who will emerge from the treatment. I am suggesting that we not shrink from this way of conceiving some of the practical problems we will face as we devise new ways to cope with various medical conditions and learn that they will induce profound psychological change. We may be offering patients a form of nonbiological death that is, nonetheless, preferable to the alternatives. I hope there will be an understanding that this prospect need not, in every case, be viewed in a negative light. There is also the positive light that is shed by an overall conception of personal identity as a practical matter—as something that is *never* a metaphysical or biological given in any case but is *always* a product of effort and will, insofar as it must always be actively chosen and maintained in the face of alternatives, for the sake of something worth doing.

NOTES

1. Two central texts in the literature that rely heavily on thought experiments are Derek Parfit's *Reasons and Persons* (1984) and Peter Unger's *Identity, Consciousness and Value* (1990).

2. The following arguments are distilled from my book-length argument in *The Bounds of Agency: An Essay in Revisionary Metaphysics* (Rovane 1998).

3. The following remarks on why this criterion of personhood is properly regarded as an ethical criterion are consistent with my account in *The Bounds of Agency.* But they differ in some detail, mainly because I am leaving out—simply for lack of space—all references to an ethical notion called *agency-regard* that I developed in that book.

4. Kant and other rationalists have tried to show that we cannot operate in this normative space of rationality without also recognizing certain moral obligations. I make no such assumption here. It is morally wrong, in Kant's view, to lie. But a lie is clearly an effort at rational influence that aims at rational response. Liars cannot expect to be believed if their interlocutors have evidence that they are lying. So, in general, liars will direct their lies at persons who do not have such evidence. Not having such evidence, it is of course rational for their interlocutors to believe their lies. So, whatever moral defect there might be in lying, it is not the sort of faulty rationality that I describe in the text. What I am referring to here is persons' efforts to influence one another that involve kinds of rational failure that are not present in cases in which liars have good reasons to lie and their interlocutors have good reasons to believe them. Rather, these are the kinds of rational failure that persons commit because of fatigue, inattention, intrinsic human limitation, irrational desire, and so on.

5. To follow through on the point of note 4: even lies and threats appeal to a person's normative commitment to achieving overall rational unity. Their point is not to get a person to go in for rational failure, but to get a person to judge that it is best, all things considered, to believe the lie or comply with the threat. And this is so with all attempts at interpersonal engagement, in which persons treat one another specifically as persons. But all lies and threats must appeal, implicitly, to the commitment that persons have to achieve overall rational unity within themselves.

6. This anti-Lockean reasoning would automatically supply a justification for any medical intervention aimed at securing both rationality and rational unity within a human being, for insofar as this is the "natural" state of the human being, it is also the "normal" state. And that is precisely what many medical practices take to be their proper goal: the preservation and restoration of what is normal.

CHAPTER SIX

Diminished and Fractured Selves

John Perry, Ph.D.

A large part of our daily life is based on knowing what to expect from human beings, and we are amazingly good at it, given how complicated humans are. After all, my brain and your brain are about as complicated as anything the world has to offer, and they control large systems, hundreds of pounds in my case, that to a casual observer might appear completely unpredictable. Consider all of the people attending this symposium. If we traced our trajectories back over the past few weeks, we would probably find paths that went all over the United States, as well as several other countries, in an unrelated fashion. And yet these paths come together today, and each of us pretty confidently expected the others to be here.

From the moment I leave my house in the morning to go to work, if not before, my life is at the mercy of my ability to figure out what people are going to do. If I'm walking or on my bike, scores of people will drive by who are capable of killing me with their cars. And yet, when I come to corners, a glance at the driver slowing down or speeding up, making eye contact or looking the other way, is usually enough for me to decide whether to cross or stay put. I put my life on the line like this almost every day, and I've made it to my sixties. Amazing, really. I must possess a pretty good way of telling what's going on with people and predicting what they will do.

We are able to deal with people effectively because we are all in possession of one of the great intellectual accomplishments of humans, which I'll call the *person theory*. The person theory isn't a scientific theory but, rather, a large, often vague, picture of how humans work that relates inner states and aspects of brains and central nervous systems—called the *mind* in the person theory—to one another and to observable stimuli and behavior. This theory allows us to make imprecise but helpful predictions about what people will do and gives us guidance on how we ought to treat them. The theory has many virtues. But it has its limits. The person theory is based on a system of *indirect classification*. We describe the internal states of humans and predict their actions on the basis of the functions those internal states perform when a well-functioning person is connected in the normal way with the external world. The theory breaks down, both descriptively and normatively, when these conditions are not met. The four case studies we are to consider put some of these limits on display.

The key features of the person theory are *intentionality, local rationality, autonomy, identity* (see appendix at the end of the chapter), and *self*. We rely on the theory to construct identities, both of ourselves and others. Our understanding of ourselves and others appeals to these identities. When they break down, our understanding of persons, both of ourselves and others, breaks down. When this happens, we have to take a step backward and try to understand, in new ways, what is going on. I'll first describe what I take to be the key concepts of the person theory and then try to step back and analyze the case studies, in which, in various ways, the person theory breaks down.

The Basics of the Person Theory

Intentionality

Intentionality is a philosopher's term for a variety of mental states and activities that we are able to describe in terms of the world. I'm talking about some of the most ordinary and ubiquitous concepts that we use to describe people, such as belief, desire, knowledge, hope, and fear. Thinking in terms of these concepts is second nature to us, but when looked at closely, they are rather puzzling and amazing.

I assume you believe that Sacramento is the capital of California. Because you have this belief, you will act in certain ways. If asked what the capital of California is, you will move your lips and vocal chords in such a way as to emit the sound "Sacramento." The belief must be something inside you, capable of being a partial cause of neurons firing and muscles moving.

But look at how we describe or identify what happens here. We don't talk about neurons or synapses. We talk about Sacramento and California. We describe your belief by what the world must be like for it to be true. Your belief is true *if* Sacramento is the capital of California, false if some other city is. So think about it. There is something in your head, a belief. This belief has a property, being true. It has that property because of a fact about two things, Sacramento and California, that are (as we attend this symposium) thousands of miles away. Isn't this amazing? How can this be?

The missing elements, which are implicit in the way we talk about beliefs, are your *ideas*. They are in your head. At least, according to the person theory, they are in your mind, and according to modern science, the mind is the brain; or, at the very least, the states of mind are determined by the states of the brain.[1] You have an idea *of* Sacramento, and an idea *of* California, and an idea of a city's *being the capital of* a state. So the picture begins to make sense. If these ideas are related in your mind in one way, you believe Sacramento is the capital of California. If they are related in another way, you believe it is not. Your belief is true if the relation among your ideas *corresponds* to the relation between Sacramento and California.

But this just raises the question: What is involved in an idea, such as my idea of Sacramento, being *of* something? What is the relation between the aspect of my brain we call my idea of Sacramento, and Sacramento, the city thousands of miles away?

I got my idea of Sacramento when I was a kid, years before I saw the city. The connection between my idea and Sacramento went through the words in my geography textbook, the words my teacher used, the connections her words had to other people and texts and their use, and eventually to people that had some pretty direct relation with Sacramento—such as seeing it. The idea was connected with certain abilities and bits of information. I learned to find Sacramento on a map. My new "Sacramento" idea became properly associated with the ideas of "being the capital of" (I already knew that Lincoln, where I lived, was the capital of Nebraska) and "California." These connections are complicated. As complicated as they are, we find it useful to talk about the brain and its ideas in terms of the things to which they have these complicated relations.

I'm connected in much the same way to all sorts of things, including, say, Aristotle and Plato, who have been dead for a couple of thousand years. It's pretty miraculous that I can think about Aristotle and have the sort of thought I express with "I wonder if Aristotle really liked Plato all that much, and if they ever hung out together and just talked about women or sports, or if they just had a student-teacher relation that was sort of formal."

But for our purposes, the main thing to note is that these complicated connections terminate in ideas in my brain that are in some sense components of my thought, something local, in my head, and capable of combining with other states to move my muscles and produce action.

Local Rationality

The person theory doesn't just involve describing people in this sort of indirect way, based on their complicated connections to the world. We think we can predict what people will do and can explain what we ourselves do in terms of these intentional states. Let's return to the driver when I'm waiting to cross the street. I decide at a glance that he sees me and can tell that I intend to cross. I assume that he has some combination of the following desires: not to run over people, not to disobey the law, not to get a ticket or to get sued. His desires, like his beliefs, I describe not in terms of neurons and synapses and the like, but in terms of things that might or might not happen in the future. And then I make a big assumption: I assume that because of having the belief that I am a pedestrian and want to cross, and that there is a stop sign, and that I have the right of way, and that it would be against the law to interfere with my passage across the street, much less to run over and maim or kill me, and because of having the desires to not hurt me, to obey the law, to not get a ticket or get sued, the driver will remain stopped. I infer what he will do, what muscles will contract, what limbs will move, what pedals will get pushed, and what, as a result of all that, he will cause the car to do. I assume that what he does will make sense, given his beliefs and desires. I assume that he will act in such a way that his desires will be satisfied, if his beliefs are true. So I take my life in my hands and cross the street in front of a huge SUV that could easily crush me.

Our expectations are based on taking people to be rational agents. The rationality is limited. We are not like Mr. Spock on *Star Trek*. Emotions often get in the way of rationality. The processes by which we come to have beliefs, settle on goals, and make decisions may be stupid and hardly deserving of the term *rational*. A 2004 poll in Tennessee showed that about 25 percent of the voters thought that Bush was in favor of taxing the rich heavily, and about 25 percent thought that Kerry favored federal support for parochial schools. These people were not very well informed, but they were probably still *locally* rational. If they wanted to tax the rich, they would vote for Bush; if they wanted federal aid for parochial schools, they would vote for Kerry. Given the beliefs they had at that moment and that were operative in their deliberations and decisions, and given their desires, urges, appetites, goals, and the like, they would behave in a way that would bring about what they wanted, or at least improve the odds of it happening. If their beliefs were false, they would fail. If their beliefs were true, they would succeed. The beliefs, whether false or true, *closed the gap* between what they wanted and what they did. *If* the beliefs were true, the action would promote their goals.

As I said, local rationality allows for a lot of stupidity. It also allows for bizarre and inexplicable desires. But however wild or irresponsible people's desires and however misinformed their beliefs, we expect that what they do will be something that will advance at least some of their desires, given their beliefs. This fundamental expectation forms the basis for our ability to interact with other humans. Without it, we would be lost.

Respect for Autonomy

The person theory is not only a way of describing people's minds and predicting what they will do. It also has a moral side; it is a prescriptive theory as well as a descriptive one; it tells us how we should *treat* persons.

We believe that within certain broad limits, people should be allowed to act as their own beliefs and desires, their own reasons, dictate. If we want to change the way they act, we should change their reasons, and we should do so in ways that respect their intentionality and rationality and fit at some level into their own projects and goals. Sometimes this is called "respecting autonomy" or "treating people as ends in themselves." This requires that if we want to influence what they are going to do, we should work to change their beliefs and desires in ways that respect their rationality, by providing them with evidence that things may be different than they think. This isn't sufficient for respecting autonomy. Threats, for example, work by changing the structure of a person's desires, but often do not respect autonomy. The threat creates a local reason for people to do something that, without the threat, wouldn't fit into their own structure of goals and beliefs. I don't really want to hand the thief my wallet. But his threat to do me bodily harm creates a local reason for doing so.

Immanuel Kant said that moral people treat other people, at least some of the time, as ends not as means (Kant 1993). This basically means that we treat them in ways that respect their own desires and goals as legitimate and we try to help them achieve these goals, even if doing so means putting our own goals on hold. Sometimes we treat people as means to achieving our own goals, but the moral person shouldn't do this all the time. Making people do something they don't want to do, even if we do so by exploiting their intentionality and local rationality, usually doesn't respect their *autonomy*, their right to act in accordance with their own vision of the good life.

We sometimes legitimately force or induce people by threats or bribes to do something that doesn't fit with their goals. We do it "for their own good." This is justified if what we make them do or prevent them from doing serves their own, more important goals. Sometimes it is justified because it fits goals that they have had, or we are sure they will have, even if they don't have them at the moment. My child doesn't want to do his homework, doesn't see it as of any value. I'm confident that at some point in the future he will have goals that will benefit from his being well educated, or at least not a complete ignoramus. This is legitimate. But what about a college student, a more-or-less adult? Based on years of experience and the incredible wisdom that comes with getting old, I may be certain that a student will be better off taking a couple of courses in computer science or statistics than simply devoting all of her units to the humanities. I may try to persuade her of this. I may be certain that if I convince her, she will come back later and thank me. Still, it's her decision. We ought to respect the autonomy of adults, even when they are autonomously doing something they will be sorry for.

Diminished and Fractured Selves

The philosophers at this symposium were asked to comment on four cases that represent a partial breakdown of the person theory, preventing those who care about the affected people from dealing with them as fully autonomous persons. It is a partial breakdown, because people who have Alzheimer disease (AD), frontotemporal dementia (FTD), severe apathy, or steroid psychosis strike us as persons and continue in many ways to be treatable as such, with a certain amount of success (with AD, this depends on how far the disease has progressed). Yet something is wrong. Their rationality is diminished; the structure of their intentional states is skewed or incomplete; we are inclined to think we should violate their autonomy for their own sakes.

In each case, some person acts in a way that deviates from the norm or paradigm of human behavior. He seems to fall short, in one way or another, of being an autonomous person. The philosopher's contribution to understanding such cases is not medical or neurological, but consists in trying to articulate, in a helpful and analytic way, the norms from which these cases deviate, so as to connect the situations with our ordinary ways of understanding and dealing with people and to help trace the implications this deviance should have, or should not have, in the way we deal with them. But before doing this, we need to say more about the person theory.

Personal Identity and the Identities of Persons

The traditional problem of personal identity for philosophers is this: Under what conditions are person A and person B one and the same person? This can be a practical problem because we have inadequate knowledge of events. The practical problem of personal identity often arises in the judicial system. The prosecutor claims that the defendant, the person sitting in the courtroom, is the same person who committed the crime, at a different place and a different time. The problem confronting the jurors is one of knowledge, of knowing the facts; it is, as philosophers say, *epistemological*.

If the jurors had a complete video of everything that happened in all the relevant parts of the world—maybe this would require more than a video; perhaps some assemblage of hyperlinked digitized videos produced by a system of video cameras spread throughout the United States and beyond as part of some future edition of the Patriot Act—they could probably be sure of the right answer. They would just rewind the video until they got to the crime, follow the movements of the criminal on the video or linked videos covering the different regions of the world into which he wandered, and see whether the criminal ended up coming into the courtroom and sitting at the defense table.

The philosopher is more likely drawn to what might be called *metaphysical* issues, issues that may remain after all of the facts are, in some sense, known.

Suppose that, as the jury follows the career of the criminal, let's call him Roscoe, he does the following. After his criminal deed, he goes to a completely up-to-date brain science facility, where brain scientists have developed a technique for duplicating brains. The hope is that, with this technology, a person with some brain deterioration can have a new brain manufactured, made of sounder material, which will be psychologically indiscernible from the original. That is, when replaced, the new brain will give rise to the same beliefs and desires and memories and intentions as the old one; the headaches will disappear, and the once inevitable strokes won't occur, but the intentionality will be the same as before. So, Roscoe has his brain duplicated. He has his original brain and his body destroyed and the duplicate brain put into a different body, Jeff's body. Jeff has just been declared brain-dead, although his other organs are in fine shape. The criminal actually goes through this just to confuse things and make it hard for anyone to trace his movements. He swears the neurosurgeons to secrecy, but they don't cooperate.

The survivor of this operation leaves the hospital and ends up in the courtroom. He admits having memories, or at least something much like memories, of committing the crime. But his lawyer claims that the criminal actually slipped up and committed suicide. A human being is an animal, and this is a different animal, a different human. The defendant is actually Jeff, with a brain transplant. He is no more the criminal than he would be if he'd gotten the criminal's liver or heart. What we have here, the lawyer argues, is Jeff, a man who had a terrible injury and who, though saved by a miracle, has lost all of his memories and in their place has delusions of a criminal past. Jeff is to be pitied, not punished. The lawyer calls some philosophers as expert witnesses (paying them less, no doubt, than other expert witnesses charge)—including Bernard Williams, say.[2]

The prosecutor is undeterred. She also calls expert witnesses, perhaps John Locke or Sydney Shoemaker. They explain that our concept of a person is not really a concept of an animal but of a certain sort of information-action system, one that our person theory fits. These philosophers maintain that the person theory actually gives us a new concept of a continuing thing, one that conceivably could breach the bounds of bodily identity. Persons are systems that pick up information from experience, develop and sustain goals, and apply the information to achieve their goals. Such systems require a certain causal basis, and some hardware on which the relevant data are stored and the relevant programs run. Usually this is provided by a single human body. But that is not a necessary requirement. Look, the philosopher-witness may intone, we recognize the possibility of having the same person without having the same body when we talk of survival in heaven or hell, or reincarnation. These may be religious fantasies, but they show that it at least makes sense to have the same person when we don't have, in any ordinary sense, the same body or the same animal. Our criminal figured out a way of surviving the death of his body. The defendant is not Jeff, with a new brain and delusions, but Roscoe, with a new body and a duplicate brain.

Whether personal identity is just a species of animal identity or something special based on the person theory, these philosophers focus on an important aspect of the theory that goes beyond intentionality and local rationality. We do not expect people, ourselves or others, to be merely locally rational, responding, in a given situation, to the impulses and desires that arise in them at that time and to the beliefs they can pick up by looking at and listening to what is happening. Persons accumulate information about the world and retain it, and they accumulate goals, aspirations, and desires, and form long-term intentions and plans that structure their lives. They have enduring traits of character and personality. They have identities.

When a psychologist or an ordinary individual (i.e., not a philosopher) talks about the identity of a person, he or she mainly has in mind not something that could be decided by fingerprints or a driver's license picture, but an enduring structure within the person, his or her own individual combination of beliefs, goals, habits, and traits of character and personality, the pattern that, as we might say, *makes* the person who he or she is.

Of particular importance is the sense the person has of herself. What properties does this person think are true of her? Which ones are most important to her? How does she see this as fitting into a narrative of her life? A psychologist might ask a person to rank in importance the properties she takes herself to have. Which properties can she not imagine not having? Can this woman imagine being a man? Would it matter a lot? Can this philosopher imagine being an accountant? Can this neuroscientist imagine being a philosopher? Does this mother find it incomprehensible that she should not be a mother, or is motherhood an accident in her life? Would being different in these ways destroy a person's sense of who she is and fracture the narrative of her life? Or could these properties be accommodated within the basic picture of herself that the person has? The most important, basic, inalienable facts about a person are more or less what the psychologist might think of as that person's identity.

When a person who has dementia (AD or FTD, as in cases 1 and 2) shows no interest in things that have long been important to him, such as cleanliness, caregivers may be distressed in large part because it is so out of character, because the person they once knew would be so shocked and ashamed at his current behavior. How do we respect such a person's autonomy? By limiting ourselves to the intentional structure that now motivates the person? Or by supplying for him what he cannot supply for himself—commitment to his own past values? And this threatened infidelity to our selves, rather than the pain and suffering connected with the later stages of the disease, is likely to be most frightening to those of us who contemplate ending up with AD or to those who have been diagnosed with the disease but are in the early stages.

With the apathetic person in case 3 (before his treatment), we may feel he has lost his identity or lost touch with his identity, not in the strict and literal sense, for it is clearly the same person who was once energetic and is now apathetic, and he will take himself to be that person. But the person's own structure of values and beliefs and priorities seems no longer to be able to motivate action. Who the person is, is no longer a guide to what the person will do. The idea that we may be helping the apathetic person return to his true identity, in the psychologist's sense, might seem to justify so dramatic an intervention as is contemplated in the case study.

Selves and the Sense of Identity

A word we often use in connection with a person's identity is *self.* Some philosophers think of selves as rather mysterious, immaterial entities. Sometimes selves are identified with the souls of Christian theology, or the essential natures that are passed along in reincarnation, or some noumenal object that exists beyond normal space and time, outside the causal realm, and joins, in some Kantian way, with the primordial structure of reality to create the world as we know it (Kant 2001). I don't think such mysterious notions of the self are required to understand the person theory. I think that a self is just a person, thought of under the relation of identity. But that sounds mysterious enough, so let me explain.

Consider what it is to be a neighbor. A neighbor is just a person, thought of as having the relation of *living next* to some person in question. A teacher is just a person, thought of as having the relation of *teaching* to some student. A father is just a person, thought of under the relation of *being the father* of.

People play important roles in other people's lives, and we give these roles titles: neighbor, teacher, father, spouse, boss, and so forth.

But we play an important role in our own life. I have a relation to myself that I don't have to anyone else: identity. Self is to *identity*, as neighbor is to *living next door* to. It is a way we think of ourselves. The basic concept of self is not of a special kind of object, but of a special kind of concept that we each have of ourselves.

Each of us has a special way of thinking about our self, that is, thinking about the person we are, through the relation of identity. We have a *self-notion*, a concept of our self as our self. I want to say a bit about this key concept, about people's sense of who they are, of their own identity, for this is a central part of the person theory. Perhaps it's unclear what I'm looking for. Sometimes the best way to find something is to first consider a case in which it is absent, and then see what is missing.

Castañeda's War Hero

Now, a sort of paradigm case of someone who doesn't know who he is, and in that sense lacks a sense of identity and has a diminished self-concept, is someone who has amnesia. Here I am thinking of a certain kind of amnesia, which may only exist, in its most perfect and full-blown state, in fiction and in philosophical examples. This is a person who, as a result of a bump on the head, has no idea who he is. One assumes that the knowledge is somewhere still in the brain, waiting to be released by another fortuitous bump on the head, or maybe surgery, or maybe just time.

I'll use an example from the great philosopher Hector-Neri Castañeda. He imagines a soldier—call him Bill—who, having performed many brave deeds in a certain battle, is injured, loses his dog-tags, and awakens with amnesia. Not only does he not know who he is, but no one else knows who he is either. He is clearly a soldier, however, and is clearly due all the rights pertaining thereto. So he is hospitalized, cured of everything but his amnesia, and goes to Berkeley on the GI Bill. In the meantime, Bill's feats during the battle have become well-known. People don't know what became of him and assume he is dead and his body unrecovered. He is awarded many medals posthumously (Castañeda 1999).

For the time being, let's concentrate on Bill as he is lying in the hospital, not knowing who he is. Now, of course, there is a sense in which he *does* know who he is. He can say, "I am me." Suppose Bill feels a pang of hunger and sees a piece of chocolate cake on the tray in front of him. Does he wonder into whose mouth this morsel should be put to relieve *his* pang of hunger? No. He knows that he is the person who is feeling the pang of hunger and the person whose arm he can control, more or less at will, and the person who has a mouth that he can't see, right below the nose, the tip of which he can see, and he knows how to direct the fork and the cake into that mouth. He knows that he is sitting in a room, on a bed, with a window looking out onto a lawn, and maybe with a radio and some magazines on the stand beside him. So, he knows a great deal about himself. Still, compared with the rest of us, he has a diminished sense of self. He doesn't have memories from which he can construct a narrative about why he is where he is. He doesn't know what values, what commitments, what beliefs, what actions led him to this hospital room. Also, because he doesn't know his own name, he can't exploit *other people's* knowledge of who he is. He can't exploit public sources of information about himself. This is something we all rely on. If I forget my phone number, I can look it up in the directory. I find out something about myself in exactly the same way that you would find out the same fact about me. Indeed, there are lots of things that make it into the public conception of us that we can't discover in any other way.

In contrast, all (or almost all) of the knowledge Bill has about himself in the hospital he acquires by what I will call, somewhat ponderously, "normally self-informative ways of knowing about a person." That is, when you see an object by holding your head erect and opening your eyes, the object you see is in front of someone. Who? You. Normally at least, this is a way of finding out what is going on in front of the person who is doing the seeing. If you feel a pang of hunger, someone is hungry, and she will have her hunger relieved if food enters her mouth and makes it to her stomach. Who? You.

Why do I say "normally"? Maybe some day brain scientists will invent a little device that will send a message from one person's eyes to another person's optic nerves, so that the second person can directly see what is front of the first. This might have some military utility. Old, frail, jittery demolition experts can guide the movements of young, healthy, steady, inexperienced ones as they defuse bombs. These experts will then have a cognitive burden that the rest of us don't have. They will have to keep track of who it is they are getting visual information from about the immediate environment. Most of us don't have to do that.

Now, consider Bill's act of extending his arm, grabbing his fork, breaking off a piece of cake, and shoving it in his mouth. I'll call that a "normally self-effecting way of acting." Moving in that way is a way that any of us can shove a piece of cake we see in front of us into our own mouths, a way of feeding ourselves. Again, "normally," because we can dream up cases in which it wouldn't work.

I'll repeat my favorite example here. At the end of Alfred Hitchcock's movie *Spellbound*, Leo G. Carroll holds a gun pointed at Ingrid Bergman, who is leaving his office, having just exposed his plot to frame his patient, Gregory Peck, for murder. We know who Carroll will shoot if he pulls the trigger: the person in front of him. Shooting a gun pointed like that is a way of shooting the person in front of you. Then we see Carroll's hand turn the gun around. The front of the gun barrel fills the whole screen. He fires. Whom does he shoot? Himself. Firing a gun held like that is a normally self-shooting way of acting. But suppose that Carroll had a donut-shaped head. Then it would be a way of shooting the person behind him. It's only a contingent fact that we don't have donut-shaped heads. That's why we need to say "normally."

So Bill, even with his amnesia, has a good deal of self-knowledge, in a perfectly reasonable sense. He proceeds to Berkeley, where he ends up getting a graduate degree in history, writing, for his dissertation, a biography of the war hero who gained his fame at the same battle from which Bill woke up with amnesia. He doesn't figure out for quite a while that he is the war hero, that his dissertation is actually autobiography.

Now the point of this is that Bill knows a great deal about a person who happens to be him. In a sense, he knows a great deal about himself, for he knows a great deal about a certain person X, and he is X. But that's not what we would ordinarily say. We would say something like this: Bill knows a great deal about the person he happens to be, but he doesn't know much about himself.

In fact, even when Bill finally figures out that it is him he is writing about, we might be reluctant to call what he is writing an autobiography. One important thing Locke emphasized was that we have a special access to our own *past* thoughts and actions. We remember them—but we can remember the past thoughts and actions of others, too. I can remember that Elwood used to think that poison oak was edible; I can remember the time Elwood ate some poison oak.

But in the case of my own thought and action, I not only remember that someone did something or that someone thought something, but also remember thinking and doing things. Shoemaker calls this "remembering from the inside." Our access to our own past thoughts and actions is phenomenologically and logically different from our memories about what others have thought and done. Remembering what one did and thought isn't *like* remembering what someone else thought and felt. And in the case of others, there is always the question of *whom*? I remember someone eating poison oak, but was it Elwood? But if I remember eating poison oak, it was me that was doing the eating.

Once Bill figures out that he is the war hero, he can assimilate all the facts he has learned about his own past into his own self-notion, his own conception of who he is. But he still won't be related to these things in the normal way, the way we expect of an autobiographer. He will know that he did these things, but he won't remember doing them.

Also, with respect to the future, I can know what you are doing, what you intend to do, what you will do. But when I know what I am doing, what I am trying to do, what I intend to do, and, in those ways, what I will do, it is based on a different way of knowing, a way each of us knows something of our own future.

A case like Bill's is pretty fantastic, but the underlying moral is generally applicable. It is a fact about the complex informational world we live in that we have lots of ways of getting information about ourselves that are not normally self-informative. The notion that Bill was able to have of himself, even when he didn't know who he was, was his *self-notion*. Self-knowledge, in the ordinary sense, is knowledge of ourselves attached to our self-notion. Knowing facts about the person you happen to be, as Bill did when he wrote his dissertation, isn't enough.

If we know who we are, if we know our own names, we can incorporate what others notice and know about us into our self-conception. We do this all the time, and in fact most of us are very concerned about what we might call our *public identities*. This is the shared conception of us that others have. It is what our mothers and fathers and sons and daughters and colleagues and bosses and employees think of us. It is what is written next to our names in the newspaper or the college catalog or in the vita on our web page. For many issues, it is a better source of information about us than any normally self-informative method of knowing.

In fact, for many of us, perhaps for most of us, some important building blocks of our own identity, our own self-conception, come from the outside, from assimilation into the *I* of the *me*—that is, by adopting as part of our self-notion opinions about ourselves that originated with the insights, or mistakes, of others. My parents tell me that I am like my grandfather, that I am a thinker not a doer, and this becomes part of my self-conception.

As we construct our public identities, we rely on the help of others. Public identities are a bit like works of art or publications; they are accomplishments that take on a life of their own.[3] And, of course, they need not be unique. I may be one person in the eyes of my surviving cousins, who meet every so often in Nebraska and reminisce about our grandmother and grandfather, uncles and aunts, and parents and one another, and a somewhat different person in the eyes of my colleagues, and so forth. My self-conception, the picture of myself that animates me and explains how I act and react, may change subtly, or not so subtly, in different situations.

So I have a sense of my own identity. Here we see this other use of *identity*. What is my identity? It is my own self-concept, the things I think hold true of me. A lot of this information I get from present perception: I think I am sitting in a chair, typing on a laptop, listening to Dixieland music, looking out the window at a rainy day. Some of it I have from memory. And some of it I have from what others have told me about myself and from applying general information about people to myself.

To reiterate:

- Each person has a special, dedicated notion, a self-notion. This notion collects information acquired in normally self-informative ways, knowledge about the person's own mental and bodily states, what the world around him is like, and what he has thought and done in the past and will do, or at least plans to do, in the future.
- Our self-notions also serve to collect information we get about ourselves in other ways, as long as we recognize that it is ourselves that the information is about. I read in the email notice of the conference what time I will be giving a paper, and where. I pick up information about myself under the name "John Perry," which is the same way that others get information about me.
- Normally we expect a person to have a complex self-concept, full of things that she has learned about herself in the past, both in normally self-informative ways and as a result of what others tell her about herself. We expect her desires and goals to be based not simply

on urges and needs that she has now, that she can discover by present feeling and introspection, but also on memories of the past and goals adopted in the past.

Two Pictures of the Self

Now I want to contrast two pictures of the self. One seems to be held by many philosophers and perhaps also by economists. I call this the *fully rational agent.* The other, which seems more accurate to me, I'll call the *multiply centered self.*

Mr. Spock of *Star Trek* seems to be a fully rational agent. When he is faced with a decision, he deliberates, taking into account all the goals he has and all that he believes. His desires are ordered by their importance; his beliefs by his degree of confidence in them, and that degree of confidence corresponds to the evidence he has for these beliefs. He rationally computes the best thing to do—that is, the thing that has the optimal chance of promoting his most important goals, given the beliefs in which he is most confident. In other words, Mr. Spock's self-concept forms a single, ordered, cognitive complex, and this complex motivates his actions. Emotions, of course, play no part in his decisions, because he has no emotions; if he did happen to have one, I suppose he would wait for it to subside or just ignore it.

My conception of the multiply centered self is based mainly on me, but I think it fits everyone I know somewhat better than does the fully rational agent picture. I think it is also a more useful picture for considering diminished and fractured selves.

My goals and beliefs combine into clusters or centers that vie for control of me. Which ones are operative and, among the ones that are operative, which ones are dominant is a situational thing. In part, I respond to what others expect. And my emotions play a large role, in pushing various centers forward or making them recede from consciousness and control.

Consciousness and control don't always go together. For example, I'm at a department meeting and someone says something that I interpret as a put-down. I become angry. The anger "takes over." I respond in a way that is locally rational, given my strong desire not to be put down, not to let those who put me down escape unpunished, and to see them wince and squirm. But other centers, other clusters of desires and beliefs, look on appalled. Part of me, a little voice, urges restraint. "Calm down. You can be almost certain you won't lose anything by shutting up. You're not a child on a playground. You are going to be sorry for this outburst. Etc. Etc." This part of me continues thinking, even as I am uttering angry words. It has lost control of the agency, or at least the part of it that controls the mouth that is speaking. Another cognitive cluster is in charge. Some goals that are important to me, like not appearing foolish, not alienating colleagues, not saying things that will be counterproductive to the deliberations of the department, play no role in this dominant cluster. They are present; they are motivating the ineffective little voice. But they have nothing to do with what I am saying out loud. The goals in control are ones that are not very important to me, or so I would have thought, things such as making sure people know how I feel, defending myself, and the like.

We seem to have here two different complexes of desires, beliefs, habits, and so forth, each associated with the same self-notion, producing action at the same time. One is producing verbal behavior; the other is producing thought and trying to take over the verbiage. Where am *I* in all of this? Looking back on it, I may feel like yet a third center of motivation, adjudicating between the two. But at the time, there are just the two centers.

Another familiar case, at least to me, is procrastination. One center of agency is busily making decisions about what to watch on TV and controlling the remote and the body that holds onto it, sprawled out on the couch. The other part controls a corner of active thought, saying, "Papers to grade, papers to grade." Maybe, at some point, for some unknown reason, I move toward my study with its desk full of papers. The one complex of desires and beliefs has managed to get control of my body, and the other is limited to plaintive questions that pop up unbidden in the imagination: How did Elaine get out of the jam? Did Steinbrenner fire George? Most of us, most of the time, keep some kind of order among our competing centers of agency.

It would not be correct to say that these centers are multiple selves. There is a constant activity of trying to maintain coherence and order. Self-deception plays an important role here. I misremember my outburst at the department meeting, trying to interpret it as something that served goals that actually had no role in motivating it. "I thought that these things needed to be said forcefully, to assure that other department members, less able to defend themselves, won't be treated similarly in the future." I say this first to my colleagues and eventually believe it myself.

Emotions are feelings, and often feelings that overwhelm us and divert us from single-minded rationality. Sometimes they get in the way; sometimes they save our lives. Emotions often serve to divert us from inappropriate deliberation and move us more quickly to action. Fear says, "Don't deliberate, flee." Sometimes a good idea, sometimes not.

I don't think Mr. Spock would be a good first mate. What gets him out of bed in the morning? What gets me out of bed in the morning? Part of me says, "Get up. Lots to do. Enjoy the sunrise. Get in an extra hour of work." Part of me says, "Who cares. Stay in bed." What decides? Deliberation? No, just a slow shift, not under my rational control, between caring about getting something done more than staying warm and comfortable—or, sometimes, the other way around.

The Three Self-notions of Eve

The *competing centers* model of the self postulates a single self-notion, with various centers competing for attention and control. Sloth, indifference, self-deception, and occasional bursts of self-discipline allow their coexistence.

But, although we naturally expect to find one self-notion or self-concept to a customer, there is no logical guarantee of this. That is, we expect one notion where all of the information gained in normally self-informative ways will be found, and various parts of which will govern all actions that are normally self-effecting. But there is no contradiction in having more than one self-notion.

This seems the natural way to regard cases of multiple personality. Consider, for example, the classic *Three Faces of Eve* (Thigpen and Cleckley 1992).[4] Here we seem to have three self-notions, Eve White, Eve Black, and Jane. Each self-notion stores some of the normally self-informative knowledge, and each is partially responsible for self-effecting action. But the distribution of knowledge is rather complex:

- Eve White is in charge of action most of the time, and during those times she gets knowledge about Eve gained through perception.
- Eve Black is in charge of action only occasionally, leading to episodes that surprise and confuse Eve White and lead her to seek therapy. Eve Black is also aware of "what happens to Eve White." That is, she gets information about the thought and actions of both the Eve White and Eve Black personalities. But she keeps track of what Eve White sees and thinks and does in a way separate from how she stores knowledge about what Eve Black sees and thinks and does.
- Jane is aware of everything, but only occasionally is in charge—only when the therapist makes a determined effort to get her to "come out."

Here we have one human, but three "personalities." Our ordinary expectations of how knowledge will be stored and action governed are upset. Eve's self-conception had to vary to make sense of what was expected of her in different situations and how she had to think of herself to survive in these situations. The pressure was so great that she lost the ability to sustain a single self-notion.

The person theory is good, but fragile. It's like an economic theory that is based on well-working First World economies but may fall short in explaining dysfunctional Third World economies. It's like our understanding of our computers and how to operate them, which works fine when the disk works fine, the software is sound, and the electricity steady. For most of us, a good bit of the time, we can think of our computers in terms borrowed from the person theory. The computer knows how to do this and that; it wants to please me; if I ask it to find a file, or send some mail, or download a file from the web, it will do so. Most of us operate computers with only the vaguest sense of what goes on inside, and yet everything we do depends on affecting what goes on inside. The pull-down menus with their English commands, the mouse, the keyboard, and all the wonderful *interfaces* that computer engineers have designed allow us to control the internal states in terms of their functions and their connections with external things. If I click on the bookmark that is *of* the *New York Times* crossword web page, my computer takes me there. I rely on this of-ness, this intentional approach to my computer. When things go wrong, I am lost. I turn to the computer expert. She fixes it or gives me an estimate of what it will cost to do so. If the cost is too high, I throw the computer away. I have no interest in trying to operate it in any ways other than that permitted by the intentional interface that I have learned.

But we can't throw people away. When they grow dysfunctional and we can no longer use the person theory to understand them and influence them, we have to try other ways. The person theory breaks down. What do we put in its place?

The Case Studies

Alzheimer Disease and Other Dementia

In the case of Alzheimer disease, some questions naturally arise, from different perspectives—of the patient himself, the caregivers (including family and close friends), and the person who is newly diagnosed with AD or who anticipates getting it and feels fear.

I should say that my father died of Alzheimer's, and I'm probably more influenced by my memories of that than by the case study of Mr. Jones. The cases are not dissimilar, except that my father was a good deal less pleasant at the various stages of the disease than Mr. Jones is described as being. I've thought a lot about AD since that time. What I would have to say about FTD is similar, so I won't separately discuss that case.

There are different stages of AD. In the early stages, when memory and functional ability decline but Mr. Jones is still able to recognize his family and deal with life more or less adequately—even though he is not the steady, reliable professional type that he seems to have been for most of his life—there seems little need to bring in the heavy machinery of personal identity theory or self theory to describe him. He is a person, the same person who was once a Fortune 500 executive. He has a self, probably still more integrated than many, with goals and memories stretching back for years. It is a diminished but not a fractured self.

In the later stages of AD, one may be tempted to say more radical things. Perhaps Mr. Jones, when he no longer seems to remember much of anything, is not really the same person who contracted the disease? One might use a Lockean memory theory to bolster this conclusion. Perhaps he is not really a person any more, not close enough to the ideal of being a rational agent?

I don't think either of these things is the right thing to say. The kinds of memory that Mr. Jones is missing are those that connect with verbal abilities: using names, recalling facts, and the like. He still reacts to his wife in a way that he doesn't react to others. Even if this were not so, even if he didn't discriminate between family, short-term paid caregivers, and strangers, I don't think any reasonable theory of personal identity would say we have a new person here. Because the progress of the disease is gradual, so is the loss of memory. At any point, until the last stages of the disease, there will be memory links, however attenuated and partial, that take us, step by step, to the person's past.

There is a stage in any wasting disease that affects the brain when one is inclined to say that, even if some brain function is left, we no longer have a person; we no longer have the person we knew and loved, but only the same live human being. But I think in the case of AD we have some temptation to say this early on. When the person ceases to recognize us, or even before, when she ceases to treat us the way she always did and ceases to be able to play the role in our lives that we are used to her playing, we suffer loss. And we must also often make difficult decisions that seem to violate her autonomy: taking away driving privileges, restricting walking privileges, not letting her write checks, supervising phone conversations—necessary because of those that prey on old people by phone and mail and now email—and ignoring temper tantrums. Our loss may be easier to deal with if we imagine that the person we knew has ceased to be; the violations of autonomy may be easier to stomach if we think we no longer have a functioning person.

But I don't think that, until the last stages of the disease, these descriptions are accurate. We are dealing with persons who have changed dramatically, and who, as the disease progresses, have diminished selves, and whose autonomy can be legitimately violated for their own sakes. But we are not dealing with nonpersons or different persons, in the strong sense. Their identities change, in the psychological sense, but they are the same person, diminished and changed.

One aspect of things may be especially troubling for the caregivers. I'm not sure what to say about it, but I'll try to say something.

Suppose that Mr. Jones, like some people with AD, comes to prefer a daily routine that would have appalled him at earlier stages of his life. He doesn't like to shower or get dressed and doesn't much care if he smells bad. (This is what happens to Mr. Smith, the person with FTD.) Imagine this occurs at a stage when he is still continent and still remembers his children and grandchildren, vaguely. But when they come to visit, he has no inclination to get cleaned up. He dresses so casually that the younger grandchildren are shocked.

Mrs. Jones is absolutely and correctly certain that this behavior would appall the old Mr. Jones. Indeed, in the early stages of the disease, this was one of the things that he feared most. Perhaps he even made her promise to force him to shower and look decent, even if she had to hire a strong-arm nurse and he protested vigorously.

Now she doesn't know what to do. The suffering, both physical and emotional, that Mr. Jones will endure if she does what is necessary to get him presentable is real and is hard for her to take. Yet the thought of her husband appearing so disheveled is hard for her to take, too. She thinks she is obliged to honor her promise to the earlier Mr. Jones. But sometimes she thinks he had no right to ask of her that she should coerce another human being to do what he so much hates to do, especially when that human being has such a limited realm of autonomy, having been stripped, by intervention or incapacity, of much of what he enjoyed in life.

What should Mrs. Jones do? Well, it probably depends on the specifics of the situation, maybe on the specifics of each visit—which grandchildren will be coming, and exactly how bad the strong-willed and ornery Mr. Jones is on that day. But in general, there are a couple of things to remember.

When we do things for children, "for their own sakes," what we have in mind is later stages of the same child, whose purposes will be served by having had their autonomy limited at the earlier stage. Billy may not want to practice the piano, but won't Bill the teenager be happy that he can understand and play music? There is danger in such thinking. Bill the teenager may wish he had been left alone. But the basic reasoning seems acceptable.

There is an asymmetry, when treating the present stage of a person in some way that he or she doesn't want to be treated, in doing it for the sake of a later stage and doing it for the sake of an earlier stage. Mrs. Jones needs to keep this in mind. In fact, Mr. Jones never will look back and thank her for getting him cleaned up. If she does it, she is not doing it for Mr. Jones's sake but for her own sake and for the sake of her children and grandchildren.

But this isn't the whole truth. The asymmetry isn't complete, for Mrs. Jones may feel some obligation to the earlier Mr. Jones. She may be doing it for the sake of his memory or for the sake of his public identity. She is sad that her own memories of her husband as a good husband and father and successful businessman have been blurred, if not obliterated, by the later stages of his life. She may wish that her children's memories will not be similarly affected, to the extent that she can help it, and that her grandchildren will not have this helpless disagreeable old man as their main memory of their grandfather. How does one balance these feelings with the reluctance to make a living, feeling person do things he doesn't want to do? Things that, if left undone, present no risk to him?

Suppose Mr. Jones had been a successful academic and had written a fine book that had gone through many editions and was considered a classic in its field. At some medium stage of the disease he completes a revision, which really converts the book into a hopeless mess, something that would humiliate the earlier Mr. Jones. He desperately wants it published. Mrs. Jones, we can suppose, has the power of attorney to prevent this. Should she allow it? Does she have a right not to allow it?

Here we could grant an unalloyed commitment to autonomy: well, it's *his* book. He has the right to do with it as *he* pleases. But this doesn't seem right. An analogy might be the case of moral responsibility. In some sense, I am morally responsible for all the wrongs I have committed. But I don't feel guilty about things I did as a kid, or even as a young man, given that they happened long ago and that the values, desires, weaknesses, and the like, that led me to do them no longer operate within me. I may continue to feel responsible for making amends, if damage continues.

It seems that in the same way that I have a right to release myself from guilt about wrong things done long ago—due to another identity, in the psychological sense—I also cease to have rights to positive things I have done. Do aging artists have a right to destroy the brilliant work they did when they were young? Do rich old men have the right to destroy the trusts and bequests they set up when they had a sounder identity?

In a sense, a person's public identity, the sense the world has of her, is an accomplishment. A life well lived—or lived well enough to instill pride in and provide a model for younger members of one's family and to produce, on balance, pleasant memories in family friends and associates—is a considerable accomplishment. We care greatly about such things. It is a commonplace that U.S. presidents care about the "verdict of history." This can seem odd. The verdict of history, when you come down to it, means what predicates will be attached to your name in history books after you die. Why does this matter? Perhaps for reasons that would not withstand extended philosophical analysis—but few reasons would. The fact is, such things do matter to us. For most of us there will be no history books, but there will be the memories of those who survive us, stories our children tell to our grandchildren, maybe a little money that will make someone's life easier. Because we care about such things, do our later selves, our diminished and fractured selves, have a right to destroy them simply by virtue of being later stages of the same person? Don't whatever principles that give our later selves a claim over our productions also give us some claim over our later selves' behavior? I think Mrs. Jones has good reasons not to let this happen. She shouldn't suppose that she is doing it for Mr. Jones's sake, in the way one guards a youth from indiscretion for the sake of the adult to follow. But she also shouldn't suppose that the present, diminished Mr. Jones has unbridled rights to destroy the public identity, the memories and opinions of him that others have, that his earlier self worked so hard to build up.

How does one rationally think of the possibility or probability of having

AD? Dying of AD is not pretty or pleasant, but I don't think it is the dying that is likely to provoke the most fear. It is the idea of being so diminished, so unlike the way we want to be, of being such a burden on loved ones, and of doing embarrassing things, that horrifies us. And yet, when we are at that stage, we may have enjoyable lives. My father, in a home in Harlingen, Texas, spent a year in which he mainly thought of himself as being in Italy, where he served in World War II, in some sort of rest home, with attractive and attentive Italian nurses. It was not an unpleasant life, but it was one he would have looked forward to with horror had he known of it in advance.

It is rational to fear the inconvenience and hardship we will cause our loved ones, the drain we may put on their energies and bank accounts, the embarrassment we may cause. And we can reasonably fear the damage that will be done to our public identity, the memories people have of us, and the record of accomplishment for which we have striven. The fact that the person doing the damage may be oneself doesn't diminish, but augments, the fear. Still, these things can be handled, with proper planning, if we are fortunate enough to know what is coming and have the resources to make plans. The key resources are loved ones, who, when the time comes, will feel obliged to the identity we have now, as well as to the person we will be then.

Some people fear dying. Hume suggested that they might try thinking of all the years before they were born and so didn't exist.[5] That wasn't so bad. Why should not existing after death be something to fear, if thousands of years of nonexistence before birth left no bad memories? Perhaps one should compare one's future as a person with AD similarly. It wasn't so bad being a child, with the diminished self and lack of autonomy that small children have.

Deep Brain Stimulation

In 2012, Charles Garrison has deep brain stimulation (DBS) to treat his apathy, a syndrome associated with his Parkinson disease. A large number of electrodes are placed in his brain. The result is a radical change, but not a change back into the sort of person he was before the onset of severe apathy. The once successful, diligent, conscientious, shy, family-oriented, quiet Mr. Garrison becomes extremely outgoing, gregarious, desirous of a great deal of attention, not very diligent at work, and only superficially interested in his family. He also becomes a Democrat and an ardent environmentalist. What happened to Mr. Garrison? One thing is clear, the second is not so clear.

Hume said that reason is, and ought to be, a slave to the passions. What exactly he meant I don't know, but I like the remark in almost any possible interpretation. The relevant one is that deliberation alone won't get us out of bed. As I lie there in bed, I may know how important the day's activities are. Perhaps I can convince myself that if I make it into work and teach my Introductory Philosophy class, young minds will be influenced that will some day save the world from tyranny and bigotry. That won't, by itself, get me out of bed. Some unfathomable something has to happen. I have to, at that moment, care about that goal. It is not enough to realize that teaching the class is what I ought to do, what I want to do, what the world wants me to do and expects me to do. Some passion, some caring, has to kick in. Dwelling on my goals, my duties, and all the good that can come from getting out of bed may give rise to such passion. Usually it does. But it doesn't constitute it. To any set of desires and beliefs that present themselves to us as our own cherished hopes and goals, we can always consistently add, "So what?" Then turn over and go back to sleep.

When someone is depressed, the "so what" addendum crops up all the time. One is tempted to think that some further beliefs or goals have been added to one's usual repertoire and changed what it is rational for one to do. All is vanity. The world is going to end before long anyway. No one really cares what I do. I can't really make a difference; when I do, I just foul things up. But that doesn't seem to be phenomenologically accurate to me. The depression comes first; the rationales for not caring come along later. The basic fact is, one just doesn't care, or doesn't care enough, to get out of bed, to grade the papers, start the coffee, or whatever. The cure, for the normal range of cases, is not more deliberation, but alarm clocks placed in the next room, automatic coffee pots, wives who give you a kick, cats that sit on your head until you get up and feed them. And antidepressants help. Without passion, or at least irritation, reason is impotent.

It seems clear that the DBS did for the more severely apathetic Mr. Garrison what my cat does for me. He wasn't depressed, according to his psychiatrist. It's hard for me to imagine what it is like to be apathetic without being depressed. Of course, there is being lazy, another thing of which I have personal knowledge. At any rate, the therapy got Mr. Garrison moving, or, given that it hasn't happened yet in my account here, will get him moving.

The means are a bit frightening, because one imagines how a technology that involves putting electrodes deep in one's brain could be exploited and abused. Slippery slopes seem all around us. If electrodes to cure apathy now, why not to cure liberalism or an inordinate love for philosophy in the near future? But the cure for slippery slopes is good shoes, sand, ropes, and things like that, not forgoing the good things they can lead us to, if we don't slip.

The DBS is frightening because it is so clearly an intervention that doesn't involve any appeal to reason, to Mr. Garrison's autonomy. But he seems to have agreed to it. There doesn't seem to be much to object to.

But was it a success? That brings up the second thing, issues that are not so clear. Mr. Garrison is no longer apathetic. He is full of energy. But is he the same Mr. Garrison? Well, in the strict and literal sense of personal identity, he surely is. Mr. Garrison changed, but he continues to exist. But where did these new needs, values, habits, commitments, and concerns come from? We wanted a quiet, diligent, Republican, family man, and got a talkative, self-absorbed, environmentalist Democrat.

We would like to have some knowledge of what happened to Mr. Garrison that ties in with the person theory. Within that theory we seem to be able to frame two alternative interpretations of what happened, and we'd like to know which one is correct.

One interpretation is this. Mr. Garrison, like all of us, was a disorderly complex of intentional states, held in some sort of equilibrium by external expectations and internal negotiations. The values exhibited in his environmentalism, Democratic Party membership, storytelling, superficial interest in his family, loquaciousness, and the like were always in there, losing the battle for dominance. This is not to say that his diligent, Republican, family-man life was a facade or a charade. That, too, was part of who he was, a part whose dominance depended on a variety of things that, for one reason or another, have ceased to hold sway. The new Mr. Garrison is one who might have emerged, under different circumstances of bodily health, for quite different reasons: using alcohol, ceasing to use alcohol, undergoing psychotherapy, finding Jesus, who knows what. The Parkinson disease, the apathy, and the therapy have shaken up the equilibrium that shaped the Mr. Garrison his family knew and wanted back. But the new Mr. Garrison is authentic, a new equilibrium among competing centers. With a little work, maybe his family can get some of the old values and cares to reemerge.

The other interpretation is that the therapy did more than reinvigorate Mr. Garrison's passion for living. It destroyed values, commitments, and traits that he had and replaced them with new ones, with no previous basis in his self. It would not necessarily follow that the therapy was a mistake. One might encourage an apathetic kid to join the Navy. One expects, indeed hopes, that his values and beliefs and approach to life will be shaken to their roots and changed. One may hope that what emerges from this experience will still like to hunt and fish and go to Giants games with his father and vote Democratic, but maybe it won't. Who knows what will happen. A vegan, Republican, Dodgers fan will still be a big win over an apathetic, depressed teenager, caring about nothing and going nowhere.

It may be that asking which sort of case we have with Mr. Garrison is a mistake. Maybe we are pressing questions that make sense within the person theory to a situation for which the theory simply doesn't apply. It is not a case beyond the limits of human understanding. But perhaps it's not a case within the limits of the human understanding that the person theory can provide, either descriptively or morally. I wish I knew.

Steroid Psychosis

There are two things about the John Fast case on which I want to comment: first, the decisions he makes, while not under the influence of steroids, to begin and then to resume taking them; then, his actions while under the influence of steroids, or while experiencing steroid psychosis.

From the point of view of a frustrated athlete, Mr. Fast's baseline life seems like a life one wouldn't want to mess with. But with a little imagination, I can relate to it. Suppose there were some pills that could make me just 10 percent mentally quicker, 10 percent better able to understand the complex arguments and subtle insights and occasional logical formulae that my philosophical colleagues throw my way. Suppose there were even some dangers. I might become 10 percent less charming and affable, 10 percent less supportive of my children. It sounds like a pretty good deal, really. I might find it attractive if I believed it would work. Not that I think I have an incredible surplus of charm and affability. But I must admit I find myself in situations in which I think I am at the limits of my intellect, or somewhat beyond, more often than I find myself at the limits of my charm and affability. With 10 percent more brainpower and 10 percent less charm and affability, I think I would be more like my philosophical heroes than I am now. Come to think of it, are you guys working on any pills like this?

But where were the desires that motivated Mr. Fast to this momentous decision over all the years when he was becoming a successful athlete, model team member, and admirable husband and father? Deep in his psyche, waiting for the right situation to provoke them, I suppose.

Why was he private about it? Because he was to some extent ashamed of his desire and his willingness to use steroids. We aren't told much about the factors that would determine whether his initial decision was rational. Steroids—at least, some of them—aren't, or didn't used to be, illegal. I gather they are often effective. Some say they are responsible for the great home-run hitters of the recent era—except for Barry Bonds, of my beloved Giants, whose late-career strength is completely due, I'm sure, to clean living and weight training. As a casual reader of sports pages, I wasn't aware that steroid psychosis is a risk. Maybe Mr. Fast wasn't either—the first time, although he certainly was the second time he decided to use steroids.

But his use, even the first time, was surreptitious. This suggests that he didn't think his reasons would bear the scrutiny of his wife and coaches or, in any case, that he didn't want to explain them. Mr. Fast perhaps had some dissatisfaction with the nature of his success, perhaps a dissatisfaction that emerged only with awareness of the possibility of taking steroids and becoming stronger. We all have desires that we don't want to disclose to others, or even fully admit to ourselves; at least I do, and I suppose others do too. So his initial decision to take steroids falls more or less within the sorts of thing we can understand with the person theory. It is motivated by beliefs that probably had some connection with reality and information available to him and by desires that were not too dissimilar to those that had motivated him throughout his life.

His later decision is harder to understand. Did he really believe that the effect of steroid use would be different the second time? Or was there something about the period when he was using steroids that we haven't grasped, that made him long to return to that condition? This would be intelligible to us with mood-enhancing drugs, drugs that produce an experiential high. The steroids seem to have produced in Mr. Fast a self-image high. Although he was, in fact, not successful on the field and was making his family miserable, it seems that he was pretty happy. But later, when he seems to be convinced that his good feelings about himself and his performance, and his attribution of jealousy to others, were a delusion, why would he want to return to his delusional state? At some level, the feelings were more important to him than reality.

This question is similar to one that arises in cases that are a sort of mirror image of Mr. Fast's: people who are more successful and even thrive when they stick to their drugs for bipolar disorder or paranoid schizophrenia, but refuse to take the drugs. In the past, at least, this was intelligible because the drugs in question can have unpleasant side effects. And people with bipolar disorder, I gather, often miss the more manic phases.

With Mr. Fast, the question that arises in my mind is whether both the initial and the subsequent decisions to use steroids were based on some deep dissatisfaction with the nature of his life and his success, a deep-seated desire to be a superstar, even if it meant losing the ties to team and family that, one would have thought, would be an incredible source of satisfaction, pride, and comfort. Was this desire so deep that he preferred to return to the state in which he seemed to himself to be such a superstar, even knowing that it was a delusion? Or so deep that, during the period between the episodes of steroid use, he didn't believe what he was told about his life as a steroid user? Or, rather, did the first period of steroid use change him in ways that make it hopeless to try to understand his subsequent decision in terms of anything like the preference structure that he had before?

We learn in ending 1 that, after resuming steroid use, John Fast believed he was a better athlete and a better person, and he liked his "new self" better than his old self. Was the good will of his teammates and family never so important to him as it seemed? Was he always a would-be prima donna, trapped in the life of a good family man and team player, with no way to get out until the steroids came along?

The person theory, as I described it, is powerful but limited. It involves an indirect way of getting at the electrical and chemical phenomena that move us. It provides no way that I know of for getting at answers to the questions of the sort I have raised here, and perhaps there are no answers.

What of Mr. Fast while taking steroids? For those of us who are limited in our interactions with other human beings to the power that the person theory gives us—who have no recourse to medical knowledge, no ability to write prescriptions—there is a paradox in dealing with people who are addicted to various things. I've had more experience than I would have chosen in dealing with people addicted to alcohol and to cocaine. The paradox is that while one is obviously dealing with someone to whom the person theory does not apply, as it should, the best thing seems to be to treat addicted individuals as autonomous humans who must take complete responsibility for their actions—to avoid becoming a "codependent."

We naturally say, in such a case, that the person isn't who he used to be. But, for reasons like those given above for the person with AD, I see no reason to deny personal identity. The problem isn't that we have a new person, but that the old person has chosen to get himself into a situation in which the character traits he once could rely on—honesty, a certain amount of prudence, feelings for those he loves—will no longer play any significant role in his decisions. But in dealing with him one must assume that memories, cares, affections, and traits of old still have some presence in his psyche. One simply has to do one's best to make sure that when the immediate effect of, say, the cocaine has abated, the person will then remember the negative effects of the cocaine use—not softened by efforts of codependents—with such horror as to produce a new equilibrium among the competing motivating complexes. Sometimes it works, and sometimes it doesn't, as the two endings to John Fast's story suggest. In such cases, as in the AD case, obligations to past and future selves can trump the mandate of autonomy.

APPENDIX: BASIC CONCEPTS OF IDENTITY AND SOME DISTINCTIONS IN TERMINOLOGY

Identity versus Similarity

The concept of *personal identity* is a special case of what is sometimes called *numeric identity*. The relevant concept of identity is expressed in various ways: "are identical," "are one and the same," and so on. If X and Y are identical, in this sense, there is just one thing that is both X and Y. So if the cows Bossie and Trixie are one and the same, if they are identical, then there is just one cow, called both "Bossie" and "Trixie." English is confusing in various ways. Almost all the words for numeric identity are also used to convey similarity. For example, imagine that we have two cows, one named "Bossie" and the other named "Trixie." They are both Guernseys, both give the same amount of milk, both are somewhat ornery when milked. We might say, "Bossie and Trixie are the same," meaning that they are similar or very much alike. Maybe the farmer liked Bossie so much that he went looking for as similar a cow as he could get; he wanted one just like Bossie. We might say he wanted the "same cow" or even "the identical cow."

Note that in the numeric sense of identity, the sense in which there is just one thing, the idea of identical twins makes no sense. If they are identical, they are not twins; if they are twins, they are not identical. *Identical* in *identical twins* doesn't mean numeric identity, but similarity, or perhaps coming from a single egg.

Logical Properties of Identity

From now on I'll use *identity* in the sense of numeric identity unless I indicate otherwise. The logical properties of identity are simply consequences of the idea of just being one thing. For example, if you have just one thing, it has all the properties it has:

- If *x* is identical with *y*, and *y* has property P, then *x* has property P (the indiscernibility of the identical).
- Further, if *x* is identical with *y*, *y* is identical with *x* (symmetry).
- If *x* is identical with *y*, and *y* is identical with *z*, then *x* is identical with *z* (transitivity).
- Everything is identical with itself; that is, for all *x*, *x* is identical with *x* (reflexivity)

Identity and Time

The Greek philosopher Heraclitus got tenure for saying that you can't step in the same river twice, because new waters are always flowing in. This is deep and profound, but not quite right. Of course you can step in the same river twice, although as you do so, you won't be stepping in exactly the same water, at least if the river is flowing at any rate at all.

If we just say that when you step in the same river at two different times, it will not be exactly similar to what it was before, this doesn't sound quite so profound. Suppose that the Cayster—Heraclitus's local river—is full of muddy water on Monday but clear on Tuesday. Then don't we have a problem? How can one river have different properties at different times, given the principle we call the indiscernibility of the identical?

We just have to be careful. The same river has the property of containing muddy water on Monday and also the property of containing clear water on Tuesday. If we include the time in the property, there is no problem.

Even if we speak in the normal-tensed way, there is no problem if we are careful. The principle of the indiscernibility of the identical implies:

- If *x* and *y* are identical, *x* has all the properties *y* has, and *x* had all the properties *y* had, and *x* will have all the properties *y* will have.

But it doesn't imply:

- If *x* and *y* are identical, *x has* all the properties *y had*, . . .

Suppose Heraclitus stands in the clear Cayster on Tuesday and says, "I stepped in this very river, the identical river, one and the same river, yesterday, and then it was muddy." From this he can infer that the river he is standing in has clear water and had muddy water the day before, and that the river he stood in yesterday had muddy water then and has clear water now. But he shouldn't have concluded that it can't be the same river he is standing in today that he was standing in yesterday.

Continuity, Causation, and Identity

The concept of identity is applied to everything—concrete objects, abstract objects (such as numbers and properties), contrived objects (such as the sequence consisting of the Eiffel Tower and Bob Dylan), clouds, wind currents, and so forth.

Persons belong to the general category of concrete things, things that have a position in space and endure through time. It is often thought that the identity conditions of concrete things amount to spatial temporal continuity. Why is the coin in my pocket now the same one I put in there this morning? Because there is a spatiotemporal continuous path that stretches from the spatiotemporal position of the coin this morning to the spatiotemporal position of the coin in my pocket now, and every point along this path is or was occupied by the coin. This is certainly something we expect of concrete objects, and it is the reason we usually think we can establish identity by establishing such a continuous history—as we imagined our jury doing in the case of Roscoe the criminal.

For most concrete things, an element of direct causality is also built into our concept. Technology provides a lot of ways of giving the illusion of a concrete thing, although what we really have is a spatiotemporally connected succession of different things, made to provide the illusion of a single thing. For example, as I type this, if I enter an *s* in the file and then go back and insert some spaces, I will think of the *s*'s I type as moving to the right along the line. This *s* isn't really a single concrete thing that is moving, but a succession of things made to give the appearance of a single thing. (Of course, it is a single *succession*, but a succession isn't a concrete thing, and a succession of *s*'s isn't an *s*). The similarity of the first *s* and the second *s* doesn't result from the usual sort of direct causality that makes a concrete thing look pretty much the same from instant to instant, even if it moves a little. Rather, one thing, *s*, is annihilated and another put in its place by the editing program. I'll call this *virtual identity*.

In the case of the succession of letters, we don't really have continuity. That would require that between any two *s*'s in the series there is another, overlapping *s*. So maybe we can distinguish between virtual identity and real identity on that basis. But, are we sure that we really have continuity in the case of ordinary objects? It isn't really something we can observe. If the scientists at the Stanford Linear Accelerator Center or the European Organization for Nuclear Research (CERN) tell me that we don't really have temporal continuity but that the careers of physical objects turn out to be full of little temporal gaps, I'll have to believe what they say. So I think we need to appeal to a concept of direct causality. The position and the characteristics of each successive stage of a physical object are explained by the position and characteristics of the earlier stage.

Ordinarily, we expect concrete things to change in gradual ways, unless there is a particular event that results in a lot of change. I expect the coin in my pocket now to look pretty much the same as the one I put in my pocket this morning. Of course, if sometime during the day I took it out and put it on a railway track and let a train flatten it, then it won't. That change can be explained, however, by the way the coin was and the pressure that the train exerted on it. The careers of concrete objects have a characteristic shape, each stage explained by the way they were and what happens to them.

This applies to humans in their physical aspects. You will expect me to look pretty much the same tomorrow as I do today, unless I get run over by a car or undergo cosmetic surgery or something like that. The similarity isn't due to some outside agency or program that is keeping track of how the successive John Perrys the world sees ought to look. It's just a consequence of the way people develop. Of course, if people look too much the same as earlier stages of themselves, when the earlier stages are considerably earlier, that also requires explanation. If the person in question lives in Los Angeles, we assume cosmetic surgery.

Our concept of the identity of a person fits into this general scheme, even though the psychological characteristics of persons—their beliefs, desires, and traits—are much different sorts of properties than the shapes and sizes and appearances of (merely) physical things. Even if we adopt a Lockean theory of personal identity and allow that we may have the same person even if we do not have the same animal—or, as Locke puts it, allow that we can have the same person when we don't have the same man—we will not have abandoned entirely our ordinary conception of identity as grounded in the direct causation of basic similarities or explicable differences in the important properties of the object in question.[6]

NOTES

1. I argue for the former view, the *identity theory*—the mind is the brain and mental states are brain states—in *Knowledge, Possibility and Consciousness* (Perry 2001).

2. Essays by Williams, Locke, and Shoemaker on personal identity can be found in the edited volume *Personal Identity* (Perry 1974).

3. See Borges's "He and I," in his *Collected Fictions* (1999).

4. See also Sizemore's *Mind of My Own* (1989); and "Hume's Deathbed Interview with James Boswell" (Hume 1947).

5. See Hume 1947.

6. For more on these issues, see *Identity, Personal Identity, and the Self* (Perry 2002).

PART III / Neuroscientists Push Back

CHAPTER SEVEN

After Locke

Darwin, Freud, and Psychiatric Assessment

Samuel Barondes, M.D.

The conveners of the symposium organized it around a single question: When an individual's personality changes radically, as a consequence of either disease or intervention, should this individual still be treated as the same person? Along with this question came the four case studies that illustrate dramatic changes in personality due to major changes in brain structure and function.

This chapter summarizes my thoughts about the central question and the discussions during the symposium. To provide some background, I start with a few comments about determinants of personality, from my perspective as a psychiatrist and neuroscientist. I then present my reaction to the cases and to the views of the philosophers—Marya Schechtman, Carol Rovane, and John Perry.

Determinants of Personality

Personality is the characteristic pattern of thoughts, feelings, and behaviors of an individual, which are usually well established by early adulthood. Psychiatrists have accumulated a large body of evidence that individual differences in personality are greatly influenced by interactions between inherited propensities and early experiences.

The origin of the inherited propensities that shape personality differences was first explained by Charles Darwin, who provided us with a way of thinking about all biological diversity. Stated in modern terms, biological diversity originates as random genetic variations, and those variations that increase survival and reproduction become more prevalent in a population. Diversity is maintained, in part, because the fitness of particular variants depends on environmental niches, including niches in social environments. Darwin was explicitly interested in personality differences as an example of these general principles and believed that his theory of evolution provided a new foundation for psychology. His belief is currently being tested as psychiatrists and geneticists have begun to search for the gene variants that influence components of personality.

The other major figure in the development of the psychiatric view of personality is Sigmund Freud. Whereas Darwin called attention to the genetic basis of personality, Freud was interested in the childhood experiences that play a major role in shaping individual personalities. He recognized that a great deal of personal development goes on in early childhood, when language and complex conscious thought have not yet developed. In Freud's view, critical childhood experiences are interpreted and stored by nonconscious mechanisms that are less rational than those used in mature conscious thought. He also believed that even as rationality develops, these unconscious mechanisms continue to play important roles in our conduct and our conceptions of who we are, a view that has been confirmed by modern cognitive psychologists. A major consequence of our frequent lack of awareness of the reasons for our behavior is that we rationalize our actions by making up stories about them. We also make up stories to cover up inconsistencies and self-deceptions.

I mention Darwin's emphasis on innate biological variations and Freud's emphasis on irrational unconscious mechanisms to contrast them with the views of John Locke, whose writings have had a strong influence on the thinking of all three of our philosophers. Locke is, of course, famous for his belief that personality starts out as a blank slate (in his words: "white paper void of all characters") and is formed exclusively by experience—a non-Darwinian view. And Locke was a great believer in the primary role of reason in personality development—a non-Freudian view. Although Schechtman, Rovane, and

Perry are well acquainted with Darwin and Freud, their views about personality are more grounded in Locke than are mine, reflecting our different intellectual ancestries. In fact, the names Darwin and Freud are absent from all of their chapters, whereas the name Locke (or "Lockean") appears 32 times in Rovane's, 6 times in Perry's, and 3 times in Schechtman's.

With this background, I will restate a few key points about the cases.

The Cases

The first three cases—Mr. Jones with Alzheimer disease (AD), Mr. Smith with frontotemporal dementia (FTD), and Mr. Garrison with Parkinson disease—have a great deal in common. All three patients are mature men, ages 55 to 75, with cognitive and behavioral disorders that reflect severe degeneration of components of their brains. In each case, we know which parts of the patient's brain have deteriorated and the distinctive patterns of microscopic pathology. In each case, we can confidently predict that there will be a continuous progression of both brain degeneration and behavioral disability. In each case, we have no way to interrupt the degenerative process, although we can offer treatments that provide some symptomatic relief.

There are some distinguishing features, however. In case 1, Mr. Jones with AD, the main personality change is progressive mental deterioration—to the point where Mr. Jones no longer recognizes his grandchildren and no longer remembers his wife's name—caused by extensive degeneration throughout his brain. In case 2, Mr. Smith with FTD, the brain degeneration is concentrated in the right frontal and temporal lobes, and the initial behavioral abnormalities are impulsivity and poor judgment due to the loss of regulatory functions that are concentrated in these brain regions. In case 3, Mr. Garrison with Parkinson disease, the initial behavioral abnormality is apathy. This is a common feature of Parkinson's, which is caused by degeneration of nerve cells that manufacture dopamine, a brain chemical that plays a key role in motivating people and in their experience of pleasure. This case is also distinguished by its fictional ending: the development of a new set of irreversible personality changes, including a switch from being shy to being gregarious, as a result of deep brain stimulation (DBS) used to treat the apathy syndrome.

Case 4, an example of steroid (androgen)–induced psychosis, is quite different from the others. Mr. Fast is a young man, with an intact young brain, who brings his personality changes on himself by taking huge doses of androgenic steroids. Another difference is that all his drug-induced personality changes—including mania, irritability, and paranoia—are reversible. When Mr. Fast stops taking androgens he reverts to his "old self," and when he goes back to them his "new self" reappears.

My Reaction to the Cases

Unlike philosophers, who think mostly about healthy people, psychiatrists are accustomed to people with mental disorders and have almost certainly had patients like the men in the four cases. When we meet a new patient, we focus on several things, while filtering them through memories of our clinical experiences. First, we try to make a diagnosis of the nature of the disorder, relying, in part, on criteria listed in psychiatry's official diagnostic manual, the *DSM* (*Diagnostic and Statistical Manual of Mental Disorders*). As we formulate our assessment, we think about the biological and environmental causes of the patient's disorder (to the extent that we understand them), its probable natural history (Is the disorder likely to get better by itself? Get worse? Recur?), and its severity. At the same time we are thinking about what we can do to help, what treatments we can offer, what hope we can give. The underlying assumption of this process of evaluation is that the patient is in a changed state, a state of disorder and distress. But patients have so much in common with who they were in their premorbid state that, in my view, there is no compelling reason to think of them as being anything other than some version of the same person. That, in short, is my gut reaction to the "same person" question at the heart of the symposium.

I understand why this question was posed in conjunction with examples of patients with degenerative brain diseases. They are victims of pathological processes that progressively destroy brain mechanisms that we all rely on to control our behavior—pathological processes for which we do not presently have good treatments. In this sense, the patients are, to varying extents, not the same as they were, and their disease-induced differences can be expected to grow. But psychiatrists are accustomed to dealing with people who are behaving differently than they used to, so the patients in the four cases are only extreme examples—just more irrational and cognitively impaired than the average psychiatric patient.

Yet there is one aspect of the "same person" issue that psychiatrists do pay close attention to: the level of psychological, social, and occupational functioning of each patient. In the diagnostic scheme of psychiatry's *DSM*, this judgment is recorded on a scale called the Global Assessment of Functioning (GAF), with each patient rated with a score from 1 to 100. On this scale, people ranked 90 to 100 are described as "[having] superior functioning in a wide range of activities, life's problems never seem to get out of hand, is sought out by others because of his or her many positive qualities. No symptoms." At the other end of the scale, those scored 1 to 10 are described as follows: "Persistent danger of hurting self or others OR persistent inability to maintain minimal personal hygiene OR serious suicidal act with clear expectation of death." Thus, this scale is a crude way of recording the degree to which patients with mental disorders, including those with progressive brain diseases, have trouble conducting their own affairs. From this information come judgments about the degree to which an individual needs to rely on family and caretakers and to give up decision-making to trusted relatives or custodians.

It should be apparent from the descriptions of Mr. Jones, Mr. Smith, and Mr. Garrison that each would be scored on the lower end of the GAF scale and that their scores will deteriorate with progressive brain degeneration. I will not venture a guess as to the specific scores they might merit on this crude scale, which individual psychiatrists tend to calibrate based on their own range of clinical experiences. But my point in calling attention to the GAF scale is to illustrate a way of thinking about the level of functioning of each patient. Whereas I see no real benefit to an all-or-none type of judgment about whether or not a sick individual is the "same person" he used to be, I am comfortable reframing the issue by following the guidelines of the psychiatric profession and viewing each patient as falling somewhere on the continuum of this crude GAF scale.

The same considerations apply to Mr. Fast, even though his disorder is reversible rather than progressive. When Mr. Fast is taking androgens, his GAF score falls—as it does with abuse of other drugs such as alcohol or amphetamines. But in my opinion he is always John Fast, whether the sober Fast or Fast-on-steroids. Just as we treat a person with severe depression, who may be startlingly different from the way she was in the nondepressed state, as the same person but with depression, so should we treat Fast-on-steroids as the same person but with certain functional differences brought out by the chemically induced brain changes.

With these thoughts in mind, I now turn to the comments of the three philosophers.

The Views of the Three Philosophers

Maya Schechtman, Carol Rovane, and John Perry each bring a distinctive viewpoint to the discussion of the cases. But I was interested to note that they all share a fascination with a rare mental disorder, dissociative identity disorder, whose essential feature as described in the current (fourth) edition of the *DSM* is "the presence of two or more distinct identities or personality states that recurrently take control of behavior" (American Psychiatric Association 1994). Schechtman devotes much of her early remarks to this disorder, which she invokes to illustrate that a human being may show "psychological alterations so profound that we question whether what we are encountering is really a single *person.*" Rovane also mentions this disorder in her first paragraph, and Perry gets to it later in his chapter, using the term "multiple personality," which is what it was called in the third edition of *DSM* (American Psychiatric Association 1980).

I suspect that all three philosophers call attention to this disorder because it shows that the "same person" issue can come up not only in people with progressive brain degeneration or chemically induced brain changes but also in people with intact, nondrugged brains. What makes this disorder so interesting to all of us is that it is an extreme example of the different patterns of thoughts and feeling that can be lurking within the same individual. But to me, it is no more than that, no more than an extreme example of what is true, to some extent, of every person—that we may show different faces of ourselves in different contexts: shy and timid in certain social situations, bold and aggressive in others; thoughtful and rational in addressing certain problems, emotional and irrational in addressing others. And all of us may be treated as a different person when we show a different face. In my view, however, we are always the same person—replete with the inconsistencies and contingent behaviors and the unconscious motives and self-deceptions that each of us has in considerable measure.

After discussing dissociative identity disorder, Schechtman turns to Locke as the inspiration for a concept she calls *forensic personhood,* which she describes as an embodiment of the psychological capacities to act as a moral agent and to enter into binding commitments, the same sort of characteristics that are considered in the GAF scale. Like the framers of that scale, Schechtman views forensic personhood as being on a continuum. As she puts it, "the capacities associated with responsibility and self-governance admit of degrees, and so there will be degrees of forensic personhood as well."

Schechtman's other main idea is the *narrative self-constitution view*, a personal autobiographical narrative that one *constitutes* (italics hers). In explaining this view, she points out its complexities. On the one hand, she believes it has some unconscious origins and "is not something someone explicitly undertakes to do." On the other hand, she views it as an active and conscious process, in that "narrative is a matter not just of passively *knowing* one's story but of interpreting one's situation and deliberating in the light of it. Self-narration involves *shaping* one's life into a coherent story as well as *conceiving* of it as such." I agree with Schechtman's basic ideas but think that, as emphasized by Freud as well as contemporary cognitive psychologists, we are much less rational, conscious, and consistent than she thinks we are and that our shaping of our personal narrative is susceptible to self-deception.

Rovane is even more explicit in emphasizing the role of conscious choice in personal identity. Her main thesis is: "*the existence of a person is never a metaphysical or biological given but is always bound up with the exercise of effort and will*" (italics hers). She also assigns central importance to *competence* which resembles Schechtman's *forensic personhood* and some components of the GAF scale. Rovane's formulation is as follows: "ethical dimensions of personhood that . . . follow on rational agency . . . constitute important additional grounds for equating the concept of person with the concept of a rational agent." And "persons alone can exercise prudential self-concern; they alone can make and keep promises to one another; they alone can treat others with respect. Last but not least—in fact, paramount for the purpose of addressing the four cases—persons alone are *competent* in the sense at work in medicine and the law." Although conscious choice is central to her view, she is not unaware of inconsistency and promptly makes clear that "a single human being can house *multiple* rational agents." But her concept of personal identity has little respect for the unconscious mental processes that I consider to be so important. While recognizing that we are "prone to positive breaches of rationality, such as self-deception and weakness of will," she insists that "rational agents can discern when they fall short of meeting the requirements of rationality, and they can regard this as grounds for self-criticism and self-improvement. And this is all it takes to qualify as a rational agent—not the ability to be perfectly rational (which no actual agent has) but the ability to identify failures and breaches of rationality along with the ability to respond appropriately with self-criticism and efforts at self-improvement."

Perry's views are as well grounded in the classical philosophical literature as Schechtman's and Rovane's, but he is also more interested in assimilating ideas from contemporary psychology and psychiatry. As a result, his opinions about personal identity and the four cases are similar to my own.

Like Schechtman, Perry likes the idea of the self-narrative, which he calls "self-notion." But unlike Schechtman, who is mainly concerned with the active and personal process of shaping one's life into a coherent story, Perry is more comfortable with its incoherence. He also emphasizes that self-concept is greatly influenced by the opinions of others, which is supported by research by social psychologists. As he puts it, "each person has a special, dedicated notion, a self-notion . . . a complex self-concept, full of things that she has learned about herself in the past, both in normally self-informative ways and as a result of what others tell her about herself."

The result of this complex and ongoing process is what Perry calls "the multiply centered self," whose formation is far less dependent on reason and far more dependent on emotion than Rovane would have us believe. As Perry explains it, "My conception of the multiply centered self is based mainly on me, but I think it fits everyone I know somewhat better than does the fully rational agent picture . . . My goals and beliefs combine into clusters or centers that vie for control of me. Which ones are operative and, among the ones that are operative, which ones are dominant is a situational thing. In part, I respond to what others expect. And my emotions play a large role, in pushing various centers forward or making them recede from consciousness and control."

Perry finds this perspective helpful in discussing the cases. For example, he interprets Mr. Garrison's dramatic change after DBS as follows: "Mr. Garrison, like all of us, was a disorderly complex of intentional states, held in some sort of equilibrium by external expectations and internal negotiations. The values exhibited in his [newfound] environmentalism, Democratic Party membership, storytelling, superficial interest in his family, loquaciousness, and the like were always in there, losing the battle for dominance."

In formulating his view of the multiply centered self, Perry emphasizes two main psychological processes that impede the development of a single, consistent self-narrative, processes that also interested Freud. The first is the conflict between our emotional reactions that predominate at certain times and our rational deliberative ones that predominate at others. The second is our propensity to self-deception, which we use, in part, to create a semblance of personal coherence. I agree with Perry that these psychological processes play a big role in our creation of our personal identities, and Perry's emphasis on them distinguishes his view from those of Schechtman and Rovane, who think that our most important characteristics are shaped by reason and conscious choice.

Because Perry and I have similar views about the four cases around which the symposium revolved, I leave it to him to have the last word about these matters:

> We are dealing with persons who have changed dramatically, and who, as the disease progresses, have diminished selves, and whose autonomy can be legitimately violated for their own sakes. But we are not dealing with nonpersons or different persons, in the strong sense. Their identities change, in the psychological sense, but they are the same persons, diminished and changed.

CHAPTER EIGHT

The Fictional Self

Michael S. Gazzaniga, Ph.D.

What makes us human? This question has been debated for centuries—what is it that differentiates us from the rest of the animal kingdom? The philosophers have talked about "persons" as narrators of the self, as rational agents, and as social beings who can form theories about themselves and those with whom they interact.

We neuroscientists seek the human spark by looking at the mechanics of brain function, hoping to find the intersection of biology and personhood. Thus, neuroscience has tended to define our species by focusing on our cognition and language skills. That is, we humans think, control our actions with conscious thought, and communicate through language to make our thoughts known. But the more neuroscientists come to understand about the human brain, the more we want to identify the specifics of human consciousness. After all, much of the research on the human brain comes from finding similarities with other species' brains—be it the mouse or the chimpanzee. In fact, one of the major recent discoveries in cognitive neuroscience—mirror neurons—was made in monkey brains. Mirror neurons have opened the door to understanding concepts such as empathy and theory of mind (more on this later). But while the motor neurons in a monkey's brain may fire when it sees another monkey pick up a banana, the monkey has not sat down to figure this out in a laboratory setting. In other words, we may have a theory about monkeys, but monkeys don't have the same level of theory about us. There's a big difference between the monkey's instinctive learning, and even its ability to communicate, and ours. What is it about human consciousness that is unique?

The deeper we delve, the more neuroscience seems to agree with the philosophers: what makes us persons, rather than merely creatures, is our ability to create a story about ourselves, and we want that story to hang together, to make some kind of coherent sense, even if our brains have to distort our perceptions to do so. Neuroscientific research has identified a mechanism, which I call the *interpreter*, in the left hemisphere of the human brain that generates such a narrative. No other species can sit around the campfire weaving experiences into a saga of personal or cultural identity.

The human brain's interpreter was discovered over a period of years of research on "split-brain" patients. I began this work in the 1960s with Roger Sperry, examining the behavior and cognition of epileptic-surgery patients. In some epilepsy patients, the corpus callosum, the nerve tract system that connects the left and right brain hemispheres, is severed to stop seizures. Post-surgery, the patients are able to return to normal life, seizure free. In fact, they appear perfectly normal unless they are tested in specific ways, which reveal that while they can still name and describe, in a normal way, information received by their left dominant hemisphere, they can respond to test stimuli presented to their right hemisphere only through nonverbal means—they are unable to provide a spoken response.

The challenge of split-brain testing was to find ways to communicate with the nonverbal, right hemisphere, without letting the left hemisphere know what we were asking. We devised nonverbal modes of responding to questions, such as patients' pointing to answers instead of speaking. To explain how we did this, understanding a basic principle of brain organization is important: if you fix your eyes on a point—for example, on a dot—all visual information to the right of the dot travels through the visual nerve pathways to your left brain hemisphere, and everything to the left of the dot is projected exclusively to your right hemisphere.

For instance, in a "simultaneous concept" test, we presented split-brain patients with two pictures, projecting one to the right hemisphere by showing it only to the left visual field and one to the left hemisphere by showing it only to the right visual field. We then spread several other pictures in front of the patients and asked them to point to the ones related to the pictures they had seen. In one such test, we flashed a picture of a chicken claw to the left hemisphere and a picture of a snow scene to the right. When choosing the related pictures, the patient, as expected, pointed to a picture of a chicken with the right hand and to one of a shovel with the left hand. But when we asked why he chose the shovel, we got an astonishing answer from the patient's verbal, left hemisphere: "Oh, that's simple. The chicken claw goes with the chicken." Then looking down at the shovel picture in his left hand, he added, "and you need a shovel to clean out the chicken shed." Entirely in the dark about the snow scene flashed to the right hemisphere, the *left* hemisphere had immediately invented a reason for the left hand's selection of the shovel.

The interpreter constantly seeks after-the-fact explanations for our behaviors and emotions. In another test, a patient saw two words simultaneously: the word *bell* was flashed to his nonverbal, right hemisphere, and *music* to his verbal, left hemisphere. When asked to point to a picture of what he saw, he chose the bell, but when asked to explain his choice, he said: "The last time I heard any music, it was coming from bells banging away." The speaking brain concocted a plausible story for why he pointed to the bell even though some of the other pictures would more obviously represent music. The design of the experiment forces these behaviors out, and the split-brain patient naturally starts to weave a story to make sense of the data he doesn't "know" his brain has received.

These simple tests reveal a mechanism in our left hemisphere that is always trying to figure out the pattern and meaning of the events we experience. In another test, we ask subjects to guess whether a light will flash above or below a fixed visual point. The test is designed so that the light flashes 80 percent of the time above and 20 percent below the fixation point. Most humans automatically try to calculate the sequence. This is called *probability matching*, and in trying to figure it out, we get it right only about 68 percent of the time. A rat, by contrast, quickly learns that if it just hits the top button all the time, it will get 80 percent of the rewards. So the rat maximizes. If you give this test to split-brain patients, the left hemisphere—trying to ascertain patterns and meanings and taking guesses—responds correctly 68 percent of the time, like most human subjects. The right hemisphere is like the rat. Having no interpreter, the hypothesis-generating mechanism, it discerns that by making the same choice every time, it will be correct 80 percent of the time.

The powerful human drive to generate a narrative was illustrated many years ago in an elegant study by Robert Kleck and Chris Strenta (1985). To help students understand the effects of being disfigured, they placed a big glob or blemish on the students' cheeks before asking them to interview other students and note how the interview subjects reacted to them when they had this blemish. Then, before sending each student-interviewer into the room where the interviewee was waiting, the experimenters wiped the goo off, claiming they were "setting" it so that it wouldn't drip during the interview. Thus, the student-interviewers went into the room believing they had a big blemish on their cheeks, even though it had been completely removed. After conducting the interviews, every student emerged to give the same report: in effect, "I had no idea how horribly people are treated when they have a blemish or a handicap."

Next, the examiners explained that videos had been made of the students being interviewed. They played back the videos, asking the interviewers to stop the video at the points where they felt the student-interviewee was reacting negatively. Each student stopped the video within seconds: "Stop, stop it, look at the way they're looking at me there!" Their left-brain interpreter would spin out a theory, even in the absence of any evidence.

We can see how the interpreter works to describe real emotional responses, in addition to imagined ones, in one of the first split-brain patients that Roger Sperry and I studied. To see whether we could get an emotional response in the tests, we slipped a pin-up picture among the pictures of more neutral objects. We flashed a picture of a spoon to the subject's left visual field/right hemisphere and asked her what she saw. Nothing. She was completely bored. Then we flashed the nude pin-up picture to the same visual field. What do you see? Again, she responded that she saw nothing. But this time she was embarrassed. When we asked her what was the matter, she said: "Oh, you guys have got a funny machine here. You come up here, you do all these tests—it's a riot."

The evidence of more than 40 years of study points to this system in the left hemisphere as having responsibility for our narrative. But what if something goes wrong with the brain? If the information the interpreter gets is distorted or altered—by diseases that diminish cognition such as Alzheimer disease and frontotemporal dementia, or by treatments such as deep brain stimulation or split-brain surgery—the story changes, too. The interpreter will create another reality to fit the data it receives. In effect, the interpreter is not so much a rational agent as a rationalizing agent.

The way that the interpreter might be involved in crafting our personal identities gradually became clear to me as I watched my father, a physician, suffer through years of strokes that slowly destroyed his cerebral cortex. Here was an example of how stories of personal identity have to be rewritten, sometimes due to extreme circumstances, over our lifespan. How would this robust, sharp-as-a-tack authoritarian figure adapt to these cognitive and physical limitations? His overall mental capacities were intact, so I could ask him what it was like to have a stroke in his right parietal lobe or his left hemisphere, and he said, "Mike, you work with what you got."

When we are forced to rationalize and recraft throughout such changes in the information coming into our brains, the interpreter works hard to maintain a more or less constant sense of self. Another time, when I asked my father how he felt (at age 78, unable to walk unaided), he simply said: "Mike, I feel 12. I always have and always will." The interpreter works with the information it gets to spin out a reinterpretation that is consistent with the last recalled conception of who we are.

Thus, the interpreter can be said to work on the data analyst's rule: GIGO (garbage in, garbage out). Why is this? The human brain has a variety of widely distributed nonverbal systems—systems that are not accessible to the verbal, left hemisphere—that are nonetheless able to produce and control behaviors. These *behaviors* are observed and noted by the interpreter system in the left hemisphere, and if it gets information that is strange in some way, it comes up with strange ideas about the nature of the world. We see this in neurological syndromes such as anosagnosia and in the split-brain patients I described above, but it occurs in the normal brain as well. If your brain receives confusing or contradictory information, your meta-cognition doesn't say, "Look, I'm just getting confusing information here, and I'll get back to normal in a bit." Instead, your brain immediately starts to make up a story using the data at hand. This phenomenon is known in social psychology as *cognitive dissonance* (Festinger 1957). Human beings find it mentally painful to deal with contradictory or "dissonant" stimuli and therefore protect themselves by immediately rationalizing what is perceived, in a way that makes cognitive sense even though it may further distort reality.

To understand why the brain works this way, we need to stop and think about what the brain is for. Why do we have this structure? The brain is there to make decisions. It makes decisions on a second-by-second, moment-by-moment, day-to-day, week-to-week basis, and it accumulates information from all kinds of sources, putting the data into some kind of decision mechanism or decision network. For the most part, this happens outside our conscious awareness.

You can demonstrate this to yourself quickly, by simply touching your finger to the tip of your nose. Notice the sensation that the tip of your finger and the tip of your nose were touched at the same time, even though the information from your finger must go through a three-foot-long neuron and the information from your nose has to travel only inches. Somehow the two are simultaneously processed in consciousness. How does that happen?

About 50 years ago, Benjamin Libet and colleagues at the University of California, San Francisco, started looking into the timing of physiological processes and how they relate to conscious experience (Libet and Jones 1957). In conscious patients undergoing surgery for epilepsy, when an electrical stimulus was applied to the cortex, the patient did not become aware of it until 500 milliseconds later. Libet subsequently theorized that, for subjective time, we automatically refer the beginning of an event backward in time, closer to the onset of the stimulus. When he tested that idea, he found that if he applied a half-second stimulus to the cortex and then 400 milliseconds later applied a single-pulse stimulus to the skin, all subjects reported that they felt the skin stimulus first. These and other experiments led Libet to believe that the brain does not use the timing of its own firing to represent timing in the real world. The brain makes its own computations that determine *when* we experience things, just as it figures out that someone is not upside down when his head is at the bottom of our retina. Most mental processing—the perception, comprehension, and reconstruction of events—goes on in our brains before we become aware of it. The interpreter steps in to generate our sense of agency, to help us maintain our illusion that events are happening in real time and under our control.

Given these tricks the brain plays on itself, perhaps it is not surprising that the brain can be hijacked. I have had the experience myself, and I can attest that the phenomenon is robust. Jim Blascovich and Jack Loomis at the University of California, Santa Barbara, have a system in their lab for creating virtual reality (Loomis, Blascovich, and Beall 2002). After donning special goggles, you are tethered to a computer system that allows you to see the world only as the computer generates it. The lab has a cement floor, and you know where you are as you walk along. Then the experimenters, through the virtual goggle system, seem to open a hole in the floor in front of you. Suddenly your brain sees this hole in the floor, and you jump back. Your heart starts to race, you gasp for air, and you become extremely anxious. Meanwhile, the graduate students in the corner are snickering. They've hijacked another brain.

Next, they put a gangplank across the hole and tell you to walk across it. There is no way you are going to walk across that gangplank, because you are afraid you will fall into the hole. You know this is ridiculous, because you know that you are, in fact, on a cement floor at the University of California. But your interpreter is taking the information it gets from its senses and coming to a totally rational conclusion—or, at least, locally rational, as John Perry terms it—about what you should and should not be doing under the circumstances. In this sense, all reality is virtual. The brain's job is to construct a picture of the world that allows you to make decisions, the kind of decisions that allow you to live another day, long enough to produce more human beings.

Part of the interpreter's job is to tell us the lies we need to believe to remain in control. When the brain loses some part of its normal processing capabilities, it's generally not aware of it and may even deny it strenuously. Many years ago, studies conducted by Teuber, Battersby, and Bender (1960) illustrated this in World War II veterans with injuries to the central visual brain pathways. If the visual system sustains damage in the optic tract, the patient instantly becomes aware of a blind spot, or scotoma, in the visual field. But if a lesion producing a scotoma of the same size occurs in the visual cortex, where the systems that monitor input from the periphery are located, the patient goes through a period of not realizing he has a blind spot. One particular patient was a mail sorter who was sent to the doctor because he kept missorting the mail in the old pigeon-hole system. He was unaware that he had tunnel vision, because that information was not getting through to the interpreter.

Similarly, in the neurological setting, one might predict that split-brain patients would not miss their right hemisphere, and that is, in fact, what happens. Their left, speaking hemisphere is working normally, and they do not notice that their right hemisphere is no longer connected. When we interviewed two split-brain patients and asked if they thought the surgery had altered the way they looked at their lives, they answered as if nothing had changed. This was their verbal, left hemisphere speaking. Although they would look at a point and not be able to describe anything to the left of it, they would not complain about that. They experience their behavior as something emanating from their volitional selves, and the interpreter incorporates that behavior into their ongoing theory of why they behave as they do.

Even though most cognitive functions are distributed throughout the left and the right hemispheres, it's clear that conscious awareness, memories, and our reflexive knowledge of these self-defining mechanisms are tightly controlled and localized. If you disconnect them, the interpreter doesn't even know it's missing something—it doesn't notice, for example, that it should be able to see the left half of your face. The conscious awareness that you have a left side of the face to miss seems to be localized close to the areas that actually process the information (in this case, the right hemisphere), and one can look at the various sensory fields and cognitive domains and come up with similar observations. One of the things we tend to overlook in our understanding of the brain is that the vast amount of its physical mass is connections—white matter starkly visible on the standard MRI (magnetic resonance image)—given over to reporting information from local centers. Information from these connections, communicating the effects of local processing to the network, is ultimately coordinated by the system we call the interpreter.

Such neuroscience suggests that the self is a narrator—sometimes an unreliable narrator—that tries to piece together a coherent story, to find some rational unity, and to give itself a feeling of control. The self has a theory about itself, a *person theory* that applies to others as well as itself. Some suggest that the more we know about biological explanations for our choices and behavior, the harder it is to sustain a narrative in which a person is an agent. However, the fact that some of our actions may be biologically determined, or predetermined, is only part of the story. Our job as scientists is to find the neurobiology that explains not only how we understand ourselves but also how we understand others, and how we are able to predict the way others are going to act.

This brings us to the question of what role our interpreter plays in moral or ethical behavior. Cognitive neuroscientists and some philosophers have long proposed that our species has some kind of common moral spark that accounts for the fact that we generally don't go around killing, stealing from, and cheating each other. Not that this never happens, of course, but certain norms of decent behavior are common across cultures.

The question is, can we use the tools of neuroscience to determine whether there is a universal ethics? Studies using brain imaging and other modalities are starting to show that the brain lights up in unique ways when facing different kinds of moral questions. Mark Hauser and his colleagues at Harvard conducted an internet survey of many thousands of people to analyze moral behavior across 138 cultures (Hauser and Carey 1998). In one of their examples, "Denise" is a passenger on a train whose engineer has fainted, and standing on the track ahead are five people. The main track splits to the right, and Denise can direct the train onto the right track, but another person is standing in that path. Is it morally defensible for Denise to turn the train, killing the one, or should she refrain, allowing the train to kill the five? It turns out that 89 percent of respondents around the world agree that it is defensible to turn the train.

Hauser next proposes a problem facing "Frank," who is on a footbridge over the train track, with a large man beside him. If he pushes this big man onto the track, the train will stop and the five people will be saved. Is it morally permissible for Frank to shove the other man to his death? Only 11 percent of Hauser's respondents say yes. The survey results show a striking unity of response across cultures, age groups, and genders. And yet, interestingly, no one can give a coherent explanation for the respondents' decisions. Each person's interpreter, as I would put it, spins a different story as to why it responded as it did to Denise's versus Frank's problem.

Using a similar problem, Joshua Greene and Jonathan Cohen at Princeton demonstrated that emotional networks in the brain are activated if the person is *physically* interacting with another (Greene et al. 2001). These networks do not light up when the subject is contemplating saving the five lives by pulling a switch, as opposed to thinking of pushing the man into the path of the train.

Again, most of us do not go around killing each other. Why is that? What keeps the species from going off the deep end? Some of the most exciting work in neurobiology relates to how we are able to "get inside the minds" of other persons. We are inherently social creatures, whose communications are driven by empathic processes that enable us to feel the emotions of others and to simulate their mental states.

One mechanism that allows us to understand our fellow humans in this manner is what seems to be our automatic and inherent capacity to imitate the expressive behavior of others. This phenomenon has been referred to as *behavioral matching* (Bernieri and Rosenthal 1991), *automatic motor mimicry* (Bavelas et al. 1987), and the *chameleon effect* (Chartrand and Bargh 1999). It describes the tendency to unconsciously mimic the bodily postures, expressive gestures, and facial expressions of persons around us. Hatfield, Cacioppo, and Rapson (1994) further linked these mechanisms to a process they termed *emotional contagion*, whereby the unconscious mimicry of others' emotional expressiveness causes us to experience these same emotions. How can these tendencies toward synchronous behavior and vicarious experience be explained?

Nearly a decade ago, Giacomo Rizzolatti and colleagues were exploring the neural basis of reaching and grasping in the awake, behaving monkey (Gallese et al. 1996). They discovered a subset of neurons in the premotor cortex that fired not only when the monkey performed a grasping action (e.g., grabbing food) but also when the monkey observed someone else (including the experimenter) perform the same action. Rizzolatti called these activated neurons *mirror neurons*: they behaved in the same way whether the action was performed by the monkey or the monkey observed the action performed by someone else. This result was surprising, because it suggested that we understand the actions of others by mapping those actions onto our own motor system. Could there also be mirror neurons for socially relevant actions, such as smiling and frowning? And could this system provide a neural substrate for empathy?

Studies at Dartmouth looked at examples of mirroring for socially relevant actions. Using functional brain imaging (fMRI), the researchers found evidence for a mirroring system for such actions in the right hemisphere's premotor system.

In one experiment, subjects watched movies of a model smiling and frowning while the subjects' brain activity was monitored with fMRI. Subjects were asked to passively watch the movies, actively imitate the movies, or perform a motor control. The right-hemisphere premotor cortex was found to be active across all conditions, consistent with a mirroring system for facial expression (Leslie, Johnson-Frey, and Grafton 2004).

In a subsequent study, subjects were asked to either imitate the facial expression (smile when the model smiles) or perform the opposite expression (smile when the model frowns). Subjects reported that performing the opposite expression felt unnatural, and their brains bore this out: they showed increased activation in the anterior cingulate, an area normally associated with error processing. Performing congruent actions (i.e., smiling when the model smiles) led to increased activation in the nucleus accumbens, an area normally associated with reward. Together, these results suggest that we may be wired to punish social conflict and reward social congruency.

These studies also found an interesting dissociation between the left and right hemispheres. When subjects were asked to passively watch the model's face, they showed increased activation in the right premotor cortex, associated with mirroring and unconscious facial expression. When subjects were asked to actively imitate the face, bilateral activation was observed, with increased activation in the left premotor cortex. The left-hemisphere premotor system was most active when subjects were making the opposite expression. Hence, subtle unconscious imitation—of the kind seen in the chameleon effect—may be mediated by the right-hemisphere premotor system, while the left premotor system may play a special role in the conscious control of the face. Phenomena such as lying and putting on one's "game face" may be mediated by this left-hemisphere premotor system.

In these ways, we are trying to understand the processes by which our species engages in cooperative and altruistic behaviors. The discovery of common mechanisms that enable us to get inside each other's minds and experience each other's feelings sheds new light on the idea of doing unto others as you want done unto you. By the same token, behavioral mimicry and emotional contagion mediated by the mirror neuron system can also explain the darker side of social interaction, such as mass hysteria and group violence.

The new field of *social neuroscience,* as it is termed by its pioneers, is concerned with understanding what is happening in the cerebral cortex during the moment-to-moment processing of social comparisons and theory-making about ourselves and others. Functional imaging has already given us a measure of the self; in the lab, we can see specific activation patterns when the person is thinking about the self as opposed to others (as well as about the self in relation to others, as during activation of empathic processes). Now fast-forward this to the clinical setting. Imagine we are in an Alzheimer disease clinic, trying to decide where a patient is on the scale of self-referential identity. It is clear that the sense of self can be compromised by any number of diseases or circumstances. Suppose the "self system" doesn't light up, or suppose it lights up but the patient shows no behavioral sense of self. Suppose the medical community then says, "There's no sense of self; there's nothing there. We have a theory about this person in our minds, but the person doesn't have a theory about himself in his own mind." Is there a point at which we begin to withhold medical care, extraordinary treatment, antibiotics? These are the kinds of question we should be thinking about. Such circumstances in clinical care are on the horizon.

CONCLUSION

Common Threads

Hilary Bok, Ph.D., Debra J. H. Mathews, Ph.D., M.A., and Peter V. Rabins, M.D., M.P.H.

This book and the symposium that initiated it were based on two premises: (1) that bringing together experts from two different fields—philosophy and neuroscience—whose work addresses the construct of personal identity would advance our understanding of the topic and be of value to individuals in both fields; and (2) that using examples of clinical cases in which individuals underwent a change in personal identity could ground the discussions and make clear the application of the construct of personal identity to the everyday world. Here we attempt to address whether these assumptions proved correct. That is, are such cross-disciplinary interactions beneficial to individuals in both fields and are they successful in advancing our understanding of the topic? Our hope was not only to answer both questions in the affirmative but also to provide a compelling argument for the application of this approach to other topics. We first discuss some of the issues that arose as a result of cultural and definitional differences between philosophers and neuroscientists, then address some of the substantive issues on which authors, both within and between disciplines, disagreed.

Cross-cultural Issues

In reviewing the proceedings of the symposium and the chapters in this volume, we came up with several generalizations. First, a single definition of personal identity was not shared by all authors at the beginning of the symposium. This is not surprising, because differences in interpretation are common in many areas of neuroscience that study person-level phenomena ("mind," "pain," "decision"). Likewise, despite definitional differences among the philosophers, they generally agreed that it is useful to have a term that marks the difference between those beings who can reason, have a view of themselves over time, and reflect on who they are and who they want to be, and those beings who cannot. Within philosophy, that term is *person*. To most philosophers, the claim that some human beings are not persons is a statement of fact, and it is sometimes true. Philosophers agree, for example, that infants do not have the capacity to reason or to reflect, and thus, to philosophers, they are not persons. This does not directly imply any claim about how we ought to treat them (see Schechtman, chapter 4; Rovane, chapter 5).

However, some of the nonphilosophers heard the assertion that some human beings are not persons as a claim that such humans are mere objects to whom we can do whatever we want. There was thus an initial resistance to the claim that infants and individuals with severe dementia are not persons. This issue was resolved when the different uses of the term *person* were clarified.

We found general agreement that there are overlaps in meaning among the terms *personal identity, personhood,* and *self,* but this lack of definitional clarity was not fully explored and certainly not resolved. Nonetheless, disagreements among participants on definitional issues were discussed when the papers were presented, and several authors refined their definitions when submitting their chapters for publication. This supports one of the major premises of the conference and this book: that interdisciplinary contact and interchange can benefit the scholars and the fields of study under discussion.

More fruitfully, the discussions following presentation of the papers generated agreement that *personal identity requires that individuals have a conception of themselves that is enduring and that can be expressed.* While Samuel Barondes raised the question of nonconscious or unconscious contributions to personal identity, there was general agreement that personal identity requires both the ability to be self-reflective and a self-generated narrative.

Another area of convergence related to the similarity of the methods used by these scholars from two very different traditions. While no single definition of personal identity emerges from these chapters, each writer approached the topic by identifying its individual elements. Furthermore, both the philosophers and the neuroscientists framed questions so that a negative answer would eliminate an issue from further discussion and a positive answer would confirm its inclusion. The philosophers approached the issues primarily by using the technique of thought experiments, the neuroscientists by performing experiments and collecting descriptive data from "accidents of nature"—that is, injured individuals whose personal identity is impaired or destroyed by disease, injury, or an exogenous influence such as drugs.

One result of these discussions was the identification of a major difference between the two disciplines. Many of the questions that occupy neuroscientists cannot be answered by the methods of philosophers; yet, in this context, philosophers and neuroscientists appeared to be addressing the same question. To clarify the situation, philosophers must state clearly what they take themselves to be doing and what kinds of question they can answer. Absent such clear explanations, it might seem to nonphilosophers that philosophers are attempting to answer fundamental scientific questions without going to the trouble of doing actual experiments. In fact, philosophers who work on personal identity hope to *clarify* what personal identity *is*; that is, philosophers seek to identify what the *it* is that neural structures and processes might constitute, not to identify those structures and processes that constitute personal identity. Philosophers do not attempt to discover facts about neuroscience, but rather attempt to understand which of the facts that scientists might discover are relevant to personal identity and which are not. The converse is also true: scientists do not (or should not) attempt to discover philosophical truths, but rather to investigate and illuminate the neural correlates of moral behaviors and emotions.

Clarifying these distinctions at the outset is essential to mutual understanding. It is also important insofar as it helps clarify what the two disciplines might gain from each other and why collaborative work is important. As noted, philosophers who work on issues of personal identity seek to explain what, exactly, the claim that someone is "the same person" over time amounts to. Obviously, if philosophers disregard neuroscience, they might develop accounts that are unrealistic or are incompatible with the facts neuroscientists discover. But if neuroscientists who work on personal identity disregard the work of philosophers, they risk using conceptions of what it means to be a person, or to be the same person over time, uncritically. Their work may be scientifically rigorous but philosophically unsophisticated or simply off the point altogether—for example, drawing conclusions about the philosophical import of their work that do not follow from their data. Neuroscientists who take their work to be relevant to questions of personal identity but do not draw on philosophical work on this topic will, at best, recapitulate work that has already been done by philosophers and, at worst, adopt some view that philosophers have already considered and rejected as conceptually inadequate or contradictory. Viewed in this light, neuroscientists and philosophers are not trying to answer the same question by different means but, rather, are engaged in two complementary tasks, both of which are necessary for a real understanding of what it means to be a person.

Substantive Issues

While the symposium and this book seek to identify convergences and differences between the two disciplines, there were also differences within the disciplines. The three philosophers put forth views of personal identity that are in some ways quite different from one another. Marya Schechtman's account turns on our capacity to form a coherent narrative about ourselves, a narrative that we can articulate and that is both internally coherent and consistent with reality, at least to some extent. In Carol Rovane's account, by contrast, the core of personal identity is rational agency in the sense needed for interpersonal engagement. John Perry takes the self to involve various centers of thought and affect that vie with one another for control, and personal identity as the attempt to render as consistent and as unified a version as possible; in this view, identity is the conception of who we are, a conception that individuals try to make coherent and consistent with reality.

While Schechtman, Perry, and Rovane offer different accounts of exactly how our use of the capacities required for rational agency makes us persons, they do not differ greatly about what capacities someone must possess if she is to be a person. All three require that she should be an agent who can act based on reasons. All three require that she should be aware of that fact and able to explain why she does what she does, at least most of the time. And all three require that the explanations she gives should be, to some extent, integrated with one another so that her understanding of why she acts as she does does not vary wildly from one moment to the next, and that she understands this history as a coherent narrative.

The neuroscientists seemed to agree with this basic account, and they used scientific results to illuminate and enrich it. As Michael Gazzaniga writes: "The deeper we delve, the more neuroscience seems to agree with the philosophers: what makes us persons, rather than merely creatures, is our ability to create a story about ourselves, and we want that story to hang together, to make some kind of coherent sense, even if our brains have to distort our perceptions to do so. Neuroscientific research has identified a mechanism, which I call the *interpreter*, in the left hemisphere of the human brain that generates such a narrative. No other species can sit around the campfire weaving experiences into a saga of personal or cultural identity."

Some neuroscientists were skeptical of what they took to be the philosophers' emphasis on conscious choice as opposed to unconscious motivation, and they emphasized the possibility that our selves might be fragmented or irrational and our internal narrators unreliable. They also warned against construing personal identity as an all-or-nothing concept and emphasized instead both the ways in which the capacities central to personal identity might be compromised or eroded and the extent to which the dramatic collapse of these capacities is continuous with smaller failures that befall all of us in ordinary life. But the neuroscientists were generally sympathetic to the philosophers' attempt to construct accounts of personal identity in which agency and the capacity to understand oneself in narrative terms are central.

The neuroscientists' account of the ways in which the internal narrator can be unreliable raised intriguing questions for which further interdisciplinary collaboration might be particularly fruitful. As noted, all three philosophers emphasized the importance of internal coherence and consistency with reality to the person's understanding of herself and of her history. The neuroscientists, especially Gazzaniga, emphasized the extent to which, even in the absence of neurological illness or dysfunction, our internal narrator can be unreliable, providing confabulation rather than a truthful account of our own thoughts. This raises an important question that was not resolved: What counts as an accurate account of our own thoughts and reasoning? What, exactly, ought our narrative to be faithful to?

Gazzaniga describes a case from his classic studies of "split-brain" patients, individuals whose corpus callosum—the band of tissue that joins the right and left hemispheres of the brain—has been transected surgically, thus preventing direct communication between them:

> In one such test, we flashed a picture of a chicken claw to the left hemisphere and a picture of a snow scene to the right. When choosing the related pictures, the patient, as expected, pointed to a picture of a chicken with the right hand and to one of a shovel with the left hand. But when we asked why he chose the shovel, we got an astonishing answer from the patient's verbal, left hemisphere: "Oh, that's simple. The chicken claw goes with the chicken." Then looking down at the shovel picture in his left hand, he added, "and you need a shovel to clean out the chicken shed." Entirely in the dark about the snow scene flashed to the right hemisphere, the *left* hemisphere had immediately invented a reason for the left hand's selection of the shovel.

In a case like this, it seems clear that the internal narrator is confabulating—inventing a plausible-sounding reason to explain apparently inexplicable behavior. Because we know why the patient pointed to the shovel, we can see his explanation as a mere rationalization.

Compare this with a more everyday case. The doorbell rings. I get up and answer it, without particularly thinking about what I am doing. If someone asks me why I got up and went to the door, I say: because the doorbell rang, and I thought I should answer it (Anscombe 1957). Normally, we would take this answer to be accurate: that *is* why I got up and answered the door. However, it is not accurate because it accurately describes my conscious thoughts. When I heard the doorbell, I did not think to myself: "That's the doorbell; I suppose I should answer it." I just got up and went to the door, without thinking about what I was doing.

In this case, as in the case Gazzaniga describes, the internal narrator comes up with a plausible story to explain what I was doing. In this case, as in the case Gazzaniga describes, this story does not describe my actual thoughts. Yet in this case, and in many similar situations or instances involving ordinary actions performed without thinking, we normally believe that people's claims about why they do what they do can be accurate, in that their spontaneous narration accurately describes their reasons for doing what they do. However, it is not immediately clear what counts as an accurate description of the reasons for acting, or what that description is meant to be an accurate description *of*. Clarifying this point and explaining the kinds of neurological failure that can cause our narratives about ourselves to become detached from reality are important topics for future collaboration.

Summary

Based on the areas of agreement we have outlined here, we propose the following definition. Personal identity is characterized by (1) an ability to express a self-narrative that recognizes the presence of an acting individual (a self or identity), *and* (2) a constructed narrative that demonstrates intentionality, reasoned choice, and coherence. This definition, arrived at through discussion and debate by individuals representing the fields of philosophy and neuroscience, can be used as the basis for further studies of the neuroscience of personal identity, with the philosophy informing the study design and, potentially, leading to research that is both scientifically and philosophically rigorous. Further collaboration between scholars from the fields of neuroscience and philosophy around questions highlighted in this chapter, and others identified through similarly interdisciplinary explorations, can lead to an improved knowledge of how and why we understand ourselves the way we do.

References

American Psychiatric Association. 1980. *Diagnostic and Statistical Manual of Mental Disorders.* 3rd ed. Washington, DC: American Psychiatric Association.

———. 1994. *Diagnostic and Statistical Manual of Mental Disorders.* 4th ed. Washington, DC: American Psychiatric Association.

Anscombe, G. E. M. 1957. *Intention.* Oxford: Oxford University Press.

Bavelas, J. B., A. Black, C. R. Lemery, and J. Mullet. 1987. Motor mimicry as primitive empathy. In *Empathy and Its Development*, ed. N. Eisenberg and J. Strayer, 317–38. New York: Cambridge University Press.

Bernieri, F. J., and R. Rosenthal. 1991. Interpersonal coordination: Behavior matching and interactional synchrony. In *Fundamentals of Nonverbal Behavior: Studies in Emotional and Social Interaction*, ed. R. S. Feldman and B. Rimé, 401–32. New York: Cambridge University Press.

Blass, D. M., K. J. Hatanpaa, J. Brandt, V. Rao, M. Steinberg, J. C. Troncoso, and P. V. Rabins. 2004. Dementia in hippocampal sclerosis resembles frontotemporal dementia more than Alzheimer disease. *Neurology* 63 (3): 492–97.

Borges, L. 1999. He and I. In *Collected Fictions.* New York: Penguin Books.

Botvinick, M. 2004. Probing the neural basis of body ownership. *Science* 305 (5685): 782–83.

Castañeda, H.-N. 1999. *The Phenomeno-Logic of the I.* Ed. J. G. Hart and T. Kapitan. Bloomington: Indiana University Press.

Chartrand, T. L., and J. A. Bargh. 1999. The chameleon effect: The perception-behavior link and social interaction. *Journal of Personality and Social Psychology* 76 (6): 893–910.

Chisholm, R. 1976. *Person and Object.* La Salle, IL: Open Court.

Churchland, P. S. 2002. Self-representation in nervous systems. *Science* 296 (5566): 308–10.

D'Argembeau, A., F. Collette, M. Van der Linden, S. Laureys, G. Del Fiore, C. Degueldre, A. Luxen, and E. Salmon. 2005. Self-referential reflective activity and its relationship with rest: A PET study. *NeuroImage* 25 (2): 616–24.

Festinger, L. 1957. *A Theory of Cognitive Dissonance.* Oxford: Row, Peterson.

Fossati, P., S. J. Hevenor, S. J. Graham, C. Grady, M. L. Keightley, F. Craik, and H. Mayberg. 2003. In search of the emotional self: An fMRI study using positive and negative emotional words. *American Journal of Psychiatry* 160 (11): 1938–45.

Frege, G. 1997 [1892]. On *Sinn* [sense] and *Bedeutung* [reference]. Reprinted in M. Beaney, ed., *The Frege Reader.* Oxford: Blackwell.

Freud, S. 1891. *Zur Auffassung der Aphasien: eine kritische Studie.* Leipzig: F. Deuticke.

Gallese, V., L. Fadiga, L. Fogassi, and G. Rizzolatti. 1996. Action recognition in the premotor cortex. *Brain* 119 (pt. 2): 593–609.

Gazzaniga, M. S. 2005. Forty-five years of split-brain research and still going strong. *Nature Reviews Neuroscience* 6 (8): 653–59.

Geach, P. 1980 [1967]. Identity. *Review of Metaphysics* 21: 3–12. Reprinted in P. Geach, *Logic Matters.* Berkeley: University of California Press.

Greene, J. D., R. B. Sommerville, L. E. Nystrom, J. M. Darley, and J. D. Cohen. 2001. An fMRI investigation of emotional engagement in moral judgment. *Science* 293 (5537): 2105–8.

Harlow, J. M. 1868. Recovery from the passage of an iron bar through the head. *Publications of the Massachusetts Medical Society* 2: 329–46.

Hatfield, E., J. T. Cacioppo, and R. L. Rapson. 1994. *Emotional Contagion: Studies in Emotion and Social Interaction.* New York: Cambridge University Press.

Hauser, M., and S. Carey. 1998. Building a cognitive creature from a set of primitives: Evolutionary and developmental insights. In *The Evolution of Mind*, ed. D. D. Cummins and C. Allen, 51–106. New York: Oxford University Press.

Hogan, R. 1976. *Personality Theory: The Personological Tradition.* Englewood Cliffs, NJ: Prentice-Hall.

Hume, D. 1947. Hume's deathbed interview with James Boswell [appendix]. In *Dialogues on Natural Religion*, ed. N. Kemp Smith. Indianapolis: Bobbs-Merrill.

———. 1978. *A Treatise of Human Nature.* Ed. P. H. Nidditch. Oxford: Clarendon Press.

Kant, I. 1993. *Grounding for the Metaphysics of Morals.* Indianapolis: Hackett.

———. 2001. *Prolegomena to Any Future Metaphysics That Will Be Able to Come Forward as Science.* Indianapolis: Hackett.

Keenan, J. P., G. G. Gallup, and D. Falk. 2003. *The Face in the Mirror: The Search for the Origins of Consciousness.* New York: Harper-Collins.

Kelley, W. M., C. N. Macrae, C. L. Wyland, S. Caglar, S. Inati, and T. F. Heatherton. 2002. Finding the self? An event-related fMRI study. *Journal of Cognitive Neuroscience* 14 (5): 785–94.

Kircher, T. T., M. Brammer, E. Bullmore, A. Simmons, M. Bartels, and A. S. David. 2002. The neural correlates of intentional and incidental self processing. *Neuropsychologia* 40 (6): 683–92.

Kleck, R. E., and C. A. Strenta. 1985. Gender and responses to disfigurement in self and others. *Journal of Social and Clinical Psychology* 3 (3): 257–67.

Leslie, K. R., S. H. Johnson-Frey, and S. T. Grafton. 2004. Functional imaging of face and hand imitation: Towards a motor theory of empathy. *NeuroImage* 21 (2): 601–7.

Libet, B., and M. H. Jones. 1957. Delayed pain as a peripheral sensory pathway. *Science* 126 (3267): 256–57.

Locke, J. 1975. *An Essay concerning Human Understanding.* Ed. P. Nidditch. London: Clarendon Press.

Loomis J. M., J. J. Blascovich, and A. C. Beall. 2002. Immersive virtual environment technology as a methodological tool for social psychology. *Psychological Inquiry* 13 (2): 103–24.

Marsh, L. 2000. Neuropsychiatric aspects of Parkinson's disease. *Psychosomatics* 41 (1): 15–23.

McCrae, R. R., P. T. Costa Jr., F. Ostendorf, A. Angleitner, M. Hrebícková, M. D. Avia, J. Sanz, M. L. Sánchez-Bernardos, M. E. Kusdil, R. Woodfield, P. R. Saunders, and P. B. Smith. 2000. Nature over nurture: Temperament, personality, and life span development. *Journal of Personality and Social Psychology* 78 (1): 173–86.

Miller, B. L., W. W. Seeley, P. Mychack, H. J. Rosen, I. Mena, and K. Boone. 2001. Neuroanatomy of the self: Evidence from patients with frontotemporal dementia. *Neurology* 57 (5): 817–21.

Milner, B., S. Corkin, and H. L. Teuber. 1968. Further analysis of the hippocampal amnesic syndrome: 14-year follow-up study of H.M. *Neuropsychologia* 6 (3): 215–34.

Nagel, T. 1971. Brain bisection and the unity of consciousness. *Synthese* 22 (3–4): 396–413.

Newen, A., and K. Vogeley. 2003. Self-representation: Searching for a neural signature of self-consciousness. *Consciousness and Cognition* 12 (4): 529–43.

Olson, E. 1997. *The Human Animal.* Oxford: Oxford University Press.

Parfit, D. 1971. Personal identity. *Philosophical Review* 80 (1): 3–27.

———. 1984. *Reasons and Persons.* Oxford: Oxford University Press.

Perry, J. 1972. Can the self divide? *Journal of Philosophy* 69 (16): 463–88.

———, ed. 1974. *Personal Identity.* Berkeley: University of California Press.

———. 2001. *Knowledge, Possibility and Consciousness.* Cambridge: MIT Press.

———. 2002. *Identity, Personal Identity, and the Self.* Indianapolis: Hackett.

Rabins, P. V. 1994. The genesis of phantom (deenervation) hallucinations: An hypothesis. *International Journal of Geriatric Psychiatry* 9 (10): 775–77.

Rabins, P. V., B. R. Brooks, P. O'Donnell, G. D. Pearlson, P. Moberg, B. Jubelt, P. Coyle, N. Dalos, and M. F. Folstein. 1986. Structural brain correlates of emotional disorder in multiple sclerosis. *Brain* 109 (pt. 4): 585–97.

Robinson, R. G. 1998. *The Clinical Neuropsychiatry of Stroke.* New York: Cambridge University Press.

Rovane, C. 1998. *The Bounds of Agency: An Essay in Revisionary Metaphysics.* Princeton, NJ: Princeton University Press.

Sacks, O. 1984. *A Leg to Stand On.* New York: Harper & Row.

Schechtman, M. 1996. *The Constitution of Selves.* Ithaca, NY: Cornell University Press.

Schilhab, T. S. S. 2004. What mirror self-recognition in nonhumans can tell us about aspects of self. *Biology and Philosophy* 19 (1): 111–26.

Shoemaker, S. 1963. *Self-Knowledge and Self-Identity.* Ithaca, NY: Cornell University Press.

———. 1970. Persons and their pasts. *American Philosophical Quarterly* 7 (4): 269–85.

Sizemore, C. C. 1989. *Mind of My Own: The Woman Who Was Known as "Eve" Tells the Story of Her Triumph over Multiple Personality Disorder.* New York: William Morrow.

Starkstein, S. E., G. Petracca, E. Chemerinski, and J. Kremer. 2001. Syndromic validity of apathy in Alzheimer's disease. *American Journal of Psychiatry* 158 (6): 872–77.
Stuss, D. T., and M. P. Alexander. 2000. Executive functions and the frontal lobes: A conceptual view. *Psychological Research* 63 (3–4): 289–98.
Teuber, H. L., W. S. Battersby, and M. B. Bender. 1960. *Visual Field Defects after Penetrating Missile Wounds of the Brain.* Cambridge, MA: Harvard University Press.
Thigpen, C. H., and H. M. Cleckley. 1992. *The Three Faces of Eve.* Rev. ed. Kingsport, TN: Arcata Graphics.
Unger, P. 1990. *Identity, Consciousness and Value.* New York: Oxford University Press.

Index